I0019919

Jos van Schijndel

Integrated Modeling using MatLab, Simulink and COMSOL

Jos van Schijndel

Integrated Modeling using MatLab, Simulink and COMSOL

with heat, air and moisture applications for building physics and systems

VDM Verlag Dr. Müller

Impressum/Imprint (nur für Deutschland/ only for Germany)
Bibliografische Information der Deutschen Nationalbibliothek: Die Deutsche Nationalbibliothek verzeichnet diese Publikation in der Deutschen Nationalbibliografie; detaillierte bibliografische Daten sind im Internet über http://dnb.d-nb.de abrufbar.
Alle in diesem Buch genannten Marken und Produktnamen unterliegen warenzeichen-, marken- oder patentrechtlichem Schutz bzw. sind Warenzeichen oder eingetragene Warenzeichen der jeweiligen Inhaber. Die Wiedergabe von Marken, Produktnamen, Gebrauchsnamen, Handelsnamen, Warenbezeichnungen u.s.w. in diesem Werk berechtigt auch ohne besondere Kennzeichnung nicht zu der Annahme, dass solche Namen im Sinne der Warenzeichen- und Markenschutzgesetzgebung als frei zu betrachten wären und daher von jedermann benutzt werden dürften.

Coverbild: www.purestockx.com

Verlag: VDM Verlag Dr. Müller Aktiengesellschaft & Co. KG
Dudweiler Landstr. 99, 66123 Saarbrücken, Deutschland
Telefon +49 681 9100-698, Telefax +49 681 9100-988, Email: info@vdm-verlag.de
Zugl.: Eindhoven, University of Eindhoven, Diss., 2007

Herstellung in Deutschland:
Schaltungsdienst Lange o.H.G., Berlin
Books on Demand GmbH, Norderstedt
Reha GmbH, Saarbrücken
Amazon Distribution GmbH, Leipzig
ISBN: 978-3-639-10669-5

Imprint (only for USA, GB)
Bibliographic information published by the Deutsche Nationalbibliothek: The Deutsche Nationalbibliothek lists this publication in the Deutsche Nationalbibliografie; detailed bibliographic data are available in the Internet at http://dnb.d-nb.de.
Any brand names and product names mentioned in this book are subject to trademark, brand or patent protection and are trademarks or registered trademarks of their respective holders. The use of brand names, product names, common names, trade names, product descriptions etc. even without a particular marking in this works is in no way to be construed to mean that such names may be regarded as unrestricted in respect of trademark and brand protection legislation and could thus be used by anyone.

Cover image: www.purestockx.com

Publisher:
VDM Verlag Dr. Müller Aktiengesellschaft & Co. KG
Dudweiler Landstr. 99, 66123 Saarbrücken, Germany
Phone +49 681 9100-698, Fax +49 681 9100-988, Email: info@vdm-publishing.com

Printed in the U.S.A.
Printed in the U.K. by (see last page)
ISBN: 978-3-639-10669-5

CONTENTS

Contents

Contents

SUMMARY

An overall objective of our work is to improve building and systems performances in terms of durability, comfort and economics. In order to predict, improve and meet a certain set of performance requirements related to the indoor climate of buildings, the associated energy demand, the heating, venting and air conditioning systems and the durability of the building and its interior, simulation tools are indispensable.

In the field of heat, air and moisture transport in building and systems, much progress on the modeling and simulation tools has been established. However, the use of these tools in an integrated building simulation environment is still limited. Also a lot of modeling work has been done for energy related building systems, such as solar systems, heat pump systems and heat storage systems. Often, these models focus on the systems and not on the coupled problem of building and systems.

This thesis presents the development and evaluation of an integrated heat, air and moisture simulation environment for modeling and simulating dynamic heat, air and moisture processes in buildings and systems. All models are implemented in the computational software package MatLab with the use of SimuLink and Comsol. The main advantages of this approach are:

First, the simulation environment is promising in solving both time and spatial related multi-scale problems.

Second, the simulation environment facilitates flexible linking of models.

Third, the environment is transparent, so the implementation of models is relatively easy. It offers a way to further improve the usage and exchange of already developed models of involved parties.

More than 25 different heat, air and moisture related models are included in this work. Most of the models are successfully verified (by analytical solutions or by comparison with other simulation results) and/or validated (by experimental data). The use of the simulation environment regarding design problems is demonstrated with case studies.

Summary

Overall is concluded that the simulation environment is capable of solving a large range of integrated heat, air and moisture problems. Furthermore, it is promising in solving current modeling problems caused by either the difference in time constants between heating venting and air conditioning components and the building response or problems caused by the lack of building simulation tools that include 2D and 3D detail simulation capabilities.

The case studies presented in this thesis show that the simulation environment can be a very useful tool for solving performance-based design problems.

Nomenclature

a	PDE Coefficient
A	Surface [m^2]
ACD	Fraction of maximum capacity of the absorption chiller [-]
c	PDE Coefficient
c	Controllable input variables of the model
C	Heat capacitance [J/K]
CCD	Fraction of maximum capacity of the mechanical chiller [-]
CCD_{max}	Maximum electrical power of the mechanical chiller [W]
COP	Coefficient of Performance [-]
c_w	Specific heat of water [J/kgK]
D	Diffusivity (m^2/s)
d_a	PDE Coefficient
dp_s	Air saturation pressure temperature derivative [Pa/K]
dt	Time step [s]
E_{hp}	Electrical power supply of heat pump [W]
Elbal	Electricity balance [W]
ElA	Electricity needed at Academic Hospital [W]
ElC	Electrical power of the mechanical chiller [W]
Epb	Purchase price of electricity [Eur/J]
Eps	Sale price of electricity [Eur/J]
EP	Electricity profit [Eur/s]
Ekill	Wasted useful heat of all gas engines [W]
E_{solar}	Solar irradiance [W/m^2]
f	PDE Coefficient
F	PDE Coefficient
g	PDE Coefficient
G	PDE Coefficient

G13	Fraction of maximum capacity of gas engines 1- 3 [-]
G45	Fraction of maximum capacity of gas engines 4 and 5 [-]
gP	Cost of supplied gas to boilers and all gas engines [Eur/s]
g_{pB}	Gas price of the boilers [Eur/m^3]
g_{pG}	Gas price of the gas engines [Eur/m^3]
g_{max13}	Maximum gas supply to gas engines 1- 3 [m^3/s]
g_{max45}	Maximum gas supply to gas engines 4 and 5 [m^3/s]
Gr	Grasshof number [-]
H	Lower Heating Value [J/m^3]
h	Heat transfer coefficient [W/m^{2}K]
h_{st}	Enthalpy of steam at 8 bar [J/kg]
h_{cwa}	Enthalpy of water at 12 $^\circ$C [J/kg]
h_{hwa}	Enthalpy of water at 100 $^\circ$C [J/kg]
i	Non controllable input variables of the model
k	Heat pump efficiency [-]
K	Thermal conductivity [W/mK]
L	Thermal conductance [W/K]
m	Mass [kg]
MF	Mass flow [kg/s]
mlA	Hot water needed at Academic Hospital [kg/s]
msA	Steam needed at Academic Hospital [kg/s]
mgG	Gas supply to all the gas engines [m^3/s]
mgB	Gas supply to the boilers [m^3/s]
n	Outward unit normal [-]
o	Output variables of the model
p	Vapour pressure [Pa]
P13	Primary energy of gas supply of gas engines 1-3 [W]
P45	Primary energy of gas supply of gas engines 4 and 5 [W]
Pr	Prandl number [-]
p_s	Saturation vapour pressure [Pa]
q	PDE Coefficient

QACDback	Unused heat of the absorption chiller [W]
QcA	Cooling needed at Academic Hospital [W]
QcB	Cooling to ice storage [W]
QcB_{max}	Maximum cooling content of the ice storage [J]
QcBsum	State-of-charge of the ice storage [J]
QCHbal	Heat balance for switching on the absorption chiller [W]
QcS_{max}	Maximum power of the absorption chiller [W]
QhA	Heat needed for heating the Academic Hospital [W]
QBbal	Heat balance for the boilers [W]
QB	Demanded heat from the boilers [W]
QhSW	Switch of supply heat to the absorption chiller [0 or 1]
QhACD	Heat from gas engines to the absorption chiller [W]
QhCH	Heat from gas engines for central heating of the Hospital [W]
Qprim	Primary energy [W]
r	PDE Coefficient
R	PDE Coefficient
Re	Reynolds number [-]
RH	Relative humidity
t	Time [s]
T	Temperature [°C], ([-] when scaled)
Totp	Total profit [Eur/s]
u	Solution of the PDE(s), e.g. temperature, moisture content, etc.
U	U value [W/m²K]
u,v	Velocity (scaled) [-]
w	Moisture content [kg/m³]
x	Position [m]
y	Position [m]
Y	Transfer function [-]
z	Position [m]

α	PDE Coefficient
β	PDE Coefficient
β_{RH}	Vapour convection coefficient for air [kg/m^{2}sPa]
γ	PDE Coefficient
δ	Vapour conduction [s]
φ	Heat flux [W/m^2]
Φ	Heat flow [W]
Γ	PDE Coefficient
η_{E13}	Electric efficiency gas engines 1-3 [-]
η_{E45}	Electric efficiency gas engines 4 and 5 [-]
η_{h13s}	Thermal efficiency of heat from gas engines 1-3 to steam [-]
η_{h13A}	Thermal efficiency of heat from gas engines 1-3 to absorb. chiller [-]
η_{h13I}	Thermal efficiency of heat from gas engines 1-3 to inter coolers [-]
η_{h45A}	Thermal efficiency of heat from gas engines 4-5 to absorb. chiller [-]
η_{h45I}	Thermal efficiency of heat from gas engines 4-5 to inter coolers [-]
η_{hK}	Thermal efficiency boilers [-]
η_{cS}	Thermal efficiency of absorption chiller (i.e. COP) [-]
η_{cC}	Thermal efficiency of mechanical chiller (i.e. COP) [-]
η_{Epub}	Public utility mean electric efficiency from primary energy [-]
λ	Lagrange multiplier
Ω	Domain
$\delta\Omega$	Boundary of domain
θ	Moisture content (m^3 water /m^3 solid material)

Subscripts

0	initial value at t=0
a	air
b	water in TES
c	condenser
i	internal
e	external

F	moisture, dependent on temperature & moisture content
er	energy roof
concr	concrete
cv	convective
glazing	glazing
in	incoming
insul	insulation
max	maximum
out	outgoing
r	radiant
t	total
T	temperature, dependent on moisture content
x	air and radiant
v	evaporator
w	moisture, dependent on moisture content

Chapter 1

General Introduction

'And the Lord said, "Behold, they are one people, and they have all one language, and this is only the beginning of what they will do. And nothing that they propose to do will now be impossible for them.[6*] 'Therefore its name was called Babel, because there the Lord confused the language of all the earth. And from there the Lord dispersed them over the face of all the earth.[9*]

The directory of [U.S. Department of energy 2006] provides information on about 300 building software tools from over 40 countries for evaluating energy efficiency, renewable energy, and sustainability in buildings. If 'language' is replaced by 'building software tool' and 'propose to do' is interpreted as 'evaluating energy efficiency, renewable energy, and sustainability in buildings', the above-mentioned prophecy seems valid. Furthermore, coupling of stand-alone building software simulation tools by means of communication between software programs already gives quite promising results in the research area of whole building Heat, Air & Moisture (HAM) responses in relation with human comfort, energy and durability [Bartak 2002, Zhai 2005]. However, considering the ancient prophecy, this coupling strategy is perhaps not the best solution on the long term.

This dissertation is concerned with an <u>integrated</u> modeling and simulation approach for the <u>HAM</u> transport mechanisms in the area of building physics and building services using a <u>single</u> scientific computational software environment.

---

[*] Genesis 11:6,9

1.1 IMPORTANCY

Because everyone spends up to 80% of the time in buildings, the whole society benefits of good human comfort. Durability implies a long service life, which means less material usage, less embodied energy and less embodied pollution. Especially moisture threatens durability with wind driven rain, rising damp, built-in moisture, airflow driven water vapour entrance and human related indoor vapour production as main sources. Furthermore, humidity changes are very significant in warm moist regions. Often over 50% of the cooling energy is latent heat. Moisture buffering may reduce that percentage and therefore can contribute to save energy resources and to reduce CO_2 production. For all these reasons, a better knowledge of the whole building Heat, Air & Moisture (HAM) balance and its effect on the indoor environment, energy consumption for heating cooling, air (de)humidification and construction durability is really needed. This is also the driving force behind the International Energy Agency Annex 41 [IEA Annex 41 2006]. To study the effects of whole building HAM response on comfort, energy consumption and enclosure durability, computational tools are, beside measurements, indispensable. Moreover, it is widely accepted that simulation can have a major impact on the design and evaluation of building and systems performances. A lot of computer applications already exist [U.S. Department of energy 2006]. New developments in this field tend to be in integrated building design [Ellis et al. 2002]. Therefore integrated HAM models capable of covering HAM transfer between the outside, the enclosure, the indoor air and the heating, ventilation and air-conditioning (HVAC) systems, are sought-after. There is no ready-to-use simulation tool that covers all issues [Augenbroe 2002]. One option is the coupling of tools [Citherlet et al. 2001, Hensen et al. 2004, Zhai et al. 2002;2003]. Recent developments in scientific computational tools such as Matlab, MathCAD, Mathematica, Maple, FlexPDE, Modelica give cause for another option: the exclusive use of a *single* computational environment. What are the limitations if we use state-of-art scientific computational software? What are the benefits and drawbacks?

1.2 STATUS OF RESEARCH

In this status of research we provide a review on HAM simulation tools and simulation environments used in the area of Building Physics and Systems.

1.2.1 HAM simulation tools

In 2005, 14 different tools were used in an IEA Annex 41 [IEA Annex 41 2006] common exercise on simulating the dynamic interaction between the indoor climate of a room and the HAM response of the enclosure. All tools model the interaction of the indoor air and the enclosure. Beside the simulation tool of this thesis, four other HAM models are stand-alone simulation tools and have promising capabilities of simulating HVAC systems: Bsim, IDA-ICE, TRNSYS, EnergyPlus.

The energy related software tools at the Energy Tools website of the U.S. Department of Energy [U.S. Department of Energy 2006] has been used for several comparison studies. A recent overview is provided by Crawley et al. [Crawley et al. 2005]. Moreover, [Schwab et al. 2004] performed a study of the same tools in order to determine what kind of simulations each tool could perform focusing on HAM capabilities. After this, the list was narrowed down to 11 programs that might simulate whole building HAM transfer and energy consumption. From this group three tools met all formulated criteria by [Schwab et al. 2004]: the ability of simulating moisture storage in building materials, indoor climate, moisture exchange in HVAC system and access to source code. The three tools are Bsim, TRNSYS and EnergyPlus. These tools are also included in the IEA Annex 41.

[Gough 1999] reviewed tools with the focus on new techniques for building and HVAC system modeling. Of the four simulation techniques investigated, the equation-based method is most relevant for this work. The other three techniques: modal, stochastic and neural networks are not based on physical parameters and are therefore less suitable for solving design problems (see Section 1.2.3). Equation based method techniques such as Neutral Model Format and IDA solver are included in IDA-ICE and TRNSYS. [Hong et al 2000] reviewed the state-of-art (April 1998) on the development and application of computer-aided building simulation. Considering integrated building design systems, it is concluded by [Hong et al. 2000] that no public domain mature systems are available to today.

1.2.2 Integration efforts

In [Citherlet et al. 2001] four approaches on integration are identified: Stand-alone, interoperable, coupled (or linked) and integrated. First, the stand-alone approach represents stand-alone programs in relation to HAM modeling. This category is already discussed in the previous Section. Second, the interoperable approach represents sharing of information of different simulation tools but without interactive data exchange during the run-time simulation process. Recent work on this subject is provided by [Hensen et al. 2004, Yahiaoui et al. 2005]. Third, the coupled approach allows sharing of information during the run-time simulation process. Recent examples of such approaches between building energy programs and CFD are presented by [Zhai et al. 2002;2003, Djunaedy 2005]. Also in this category, [Clarke 2001] describes the approach to following domain coupling as implemented within the ESP-r system: (1) building thermal and visual domains; (2) building/HVAC and distributed fluid flow domains; (3) inter- and intra-room airflow domains; (4) construction heat and moisture flow domains. Fourth, the integrated approach represents simulation of different domains within a single simulation environment. [Citherlet et al. 2001] refer to ESP-r and EnergyPlus as examples of such approach. [Bradley et al. 2005] provide recent information on integrated applications in TRNSYS. As mentioned before, the aim of this thesis is to investigate integrated HAM models in a single simulation environment, which corresponds with the integrated approach.

1.2.3 Simulation environment requirements

Important requirements for a simulation environment are discussed now. First, the aim is to *predict dynamic HAM* processes in buildings and systems up to timescales of order of ~seconds. Therefore it is essential that the simulation environment incorporates a suitable model that is able to effectively simulate responses of indoor climates of whole buildings with the required timescale. A second aim is to integrate local HVAC and primary systems and controller models. In this case the so-called forward modeling approach [Rabl 1998] is applicable [ASHRAE 2005]. In this approach, the objective is to predict output variables of a specific model with known structure and known parameters. For this type of models, the known structure is often based on Ordinary Differential Equations (ODEs). Therefore it should be relatively easy to implement and couple ODEs models in the simulation environment. A third aim is to include 2D & 3D models, to simulate local effects in constructions and indoor climate. The structure of these geometry-based models is often based on Partial Differential Equations (PDEs). The use of PDEs should also be facilitated.

Furthermore, the modeling and simulation results should be reproducible and accessible. This also means that the relation between mathematical model and numerical implemented model should be clear.

1.2.4 The MatLab environment

The MatLab environment has promising capabilities of meeting all mentioned requirements:

(1) *A whole building (global) model* has already been developed in Matlab. This model, called HAMBase (Heat, Air and Moisture model for Building And Systems Evaluation), is developed by [de Wit 2006]. The model originates from an energy-based model ELAN [de Wit et al. 1988]. Over the years, this research model has continuously been improved and implemented in MatLab [de Wit 2006; van Schijndel et al. 1997;1999]. Important features for this work are: multizone modeling, response factors based network, fixed time step, solar & shadow calculations, multi climate.

(2) HVAC & primary systems (local) models, based on ODEs, can be implemented using SimuLink. This is a platform for multi domain simulation of dynamic systems. SimuLink is integrated with MatLab. Important features are: libraries of predefined blocks including controllers; ability to interface with other simulation programs; ordinary differential equations (ODE) modeling [Ashino, et al. 2000] that can accommodate continuous, discrete, and hybrid systems.

(3) Indoor airflow and constructions (local) models, based on PDEs, can be implemented using Comsol. This is an environment for modeling scientific and engineering applications based on partial differential equations (PDE). It offers a multiphysics modeling environment, which can simultaneously solve any combination of physics, based on the proven finite element method. Important features are: direct and iterative solvers; linear and nonlinear, stationary and time dependent analyses of models; modeling in 1D, 2D, 3D and of course interface to MatLab/SimuLink.

Related toolboxes in Matlab that are quite useful, in relation with this research, are listed below. Furthermore, because the models are developed in the same simulation environment, it should be relatively easy to couple models.

[Riederer 2005] provides a recent overview of Matlab/SimuLink based tools for building and HVAC simulation. SIMBAD [SIMBAD 2005] provides HVAC models and related utilities to perform dynamic simulation of HVAC plants and controllers. [El Khoury 2005] presents a multizone building model in SIMBAD. A similar thermal Toolkit named ASTECCA is presented in [Novak et al. 2005, Mendes 2003]. Several tools for fault detection and diagnosis for indoor climate systems are provided by [Yu 2003]. All previous mentioned models focus on thermal processes with limited capabilities for moisture transport simulation. In addition to these thermal oriented tools, this thesis presents how models that include moisture transport, can be simulated in SimuLink. The International Building Physics Toolbox (IBPT) [Weitzmann et al. 2003] is constructed for the thermal system analysis in building physics. The tool capabilities also include 1D HAM transport in building constructions and multi-zonal HAM calculations [Sasic Kalagasidis 2004]. All models including the 1D HAM transport in building constructions are implemented using the standard block library of SimuLink. The developers notice the possibility to couple to other codes / procedures for 2D and 3D HAM calculations. In addition, this dissertation shows how this can be done using Comsol.

Several Matlab toolboxes, such as the System Identification Toolbox and Neural Network Toolbox can be used for so-called data-driven approaches [Rabl 1998, ASHRAE 2005]. In this case, input and output variables are known and the objective is to estimate system parameters. The use of these toolboxes is demonstrated in the following papers. [Garcia-Sanz 1997] uses the System Identification Toolbox for simulating the thermal behaviour of a building with a central heating system including advanced controllers. [Virk et al. 1998] uses the same approach for modeling the thermal behaviour of a full scale HVAC system. [Mechaqrane 2004] uses the Neural Network toolbox to predict the indoor temperature of a residential building.

The Optimization Toolbox solves constrained and unconstrained continuous and discrete problems. The toolbox includes functions for linear programming, quadratic programming, nonlinear optimization, nonlinear least squares, nonlinear equations, multi-objective optimization, and binary integer programming. This toolbox is useful for optimizing design parameters. [Felsmann et al. 2003] used the MatLab Optimization Toolbox in combination with TRNSYS to optimize the control strategy for getting minimal costs and energy demand. [Liao et al. 2004] commissions a physical model for an existing system by optimising the model parameters.

1.2.5 Other simulation environments

As mentioned in Section 1.2.3, the availability of a whole building HAM model is regarded as essential for this research. The following scientific computational tools: MathCAD, Maple, Mathematica and FlexPDE, all lack such a model. A further review on these tools is therefore omitted.

The simulation environment Modelica however, contains a thermal building model. Modelica is an object-oriented language, suited for multi-domain modeling. To simulate a Modelica model, a so-called translator is needed to transform the Modelica model into the appropriate simulation environments. Dymola [Olsson 2005] is such a Modelica translator including interfaces to MatLab and SimuLink. [Pohl 2005] developed a simulation management tool in MatLab to provide easy and efficient access. [Felgner et al. 2002] presents a model library containing: Building, weather, heating and controller. The building models were verified by TRNSYS. [Nytsch-Geusen et al. 2005] present a hygrothermal model. A model of a hygrothermal wall is implemented successfully. The aim is to implement room models as well as models for windows, air volumes and inhabitants. [Casas et al. 2005] presents a Modelica model for the simulation of air dehumidification by means of a desiccant wheel. [Saldami et al. 2005] presents recent developments of modeling PDEs with Modelica. The authors verified their results with Comsol. In addition to work done in Modelica this thesis will also include models for indoor airflow and HAM transport in constructions.

1.3 PROBLEM STATEMENT

All HAM simulation tools mentioned in the previous Section face at least one limitation that cannot be solved by the tool itself. Either a problem occurs at the integration of HVAC systems models into whole building models. A main problem is caused by the difference in time constants between HVAC components and controllers (order of seconds) and building response (order of hours) [Clarke 2001]. This can cause long simulation times [Gouda et al. 2003, Felsman et al. 2003]. Or, a problem occurs at the integration of 2D and 3D geometry based models (for example airflow and HAM response of constructions) into whole building models. The problem is caused by the lack of lumped parameter tools that include internal 2D, 3D finite element method (FEM) capabilities [Sahlin et al. 2004]. The aim of this work is to confront the Matlab/SimuLink/Comsol simulation environment with the above-mentioned problems. This brings us to the next general questions:

(I) *How* can the Matlab/SimuLink/Comsol simulation environment contribute in solving these common modeling problems?

(II) What is the *use* of this simulation environment for design?

1.4 OBJECTIVES AND METHODOLOGY

This Section provides the *overall* objectives and methodology of the thesis based on the previously mentioned general problems I and II. Further on, each chapter presents it's own objectives and methodologies such as literature review, modeling methods, validation methods, application and evaluation.

1.4.1 Research

The research oriented *objectives* to cover problem I are:

(i)   the *development* of an integrated HAM modeling and simulation environment including HVAC systems models and 2D/3D geometry based models.

(ii)   *verification and validation* (V&V) of the developed models by current practice. (The reader should notice that this is a problem on it's own, especially for non linear systems with large degrees of freedom (DOF) such as computational fluid dynamics (CFD). The pioneers in the development of methodology and procedures in validation of PDE-based models, can be found in this research area. [Oberkampf & Trucano 2002] provide an overview of the fundamental issues in verifying and validating CFD. They conclude that 'to achieve the level of maturity in CFD ... and most analyses are done without supporting comparisons will require a much deeper understanding of mathematics, physics, computation, experiment, and their relationships than is reflected in current V&V practice').

(iii)   *evaluation* of the simulation environment in terms of: accessibility and repeatability of the modeling and simulation results, limitations, drawbacks and benefits.

The *methodology* used in developing an integrated HAM modeling and simulation environment was to implement the required models, as mentioned in Section 1.2.4, into SimuLink. The three steps were:

(1) *Implementing a whole building model.* Starting point was the already developed (global) whole building model HAMBase [de Wit 2006] in MatLab. In order to integrate the HAMBase model into SimuLink, the discrete model was transformed into a continuous model HAMBase_S [de Wit 2006]

(2) *Implementing HVAC and primary systems models.* Local HVAC and primary systems based on ODEs were integrated into SimuLink by the use of S-Functions.

(3) *Implementing airflow and construction models.* This required two steps. First, it had to be shown that PDE based local airflow and construction models could be accurately implemented and simulated by Comsol. Second, this type of models had to be integrated into SimuLink.

1.4.2 Design

Simulation can be used to predict building and system performances and confront the results with fixed criteria before actually built. It is widely accepted that the quality of designs can be verified and improved by the proper use of modeling and simulation tools. Verification of the design quality can be achieved by simulating the (predicted) performance and showing that it satisfies the demanded performance. Improvement of the quality is often expressed using classifications. For example, [ASHRAE 2003] gives an overview of the classification of the climate control potential in buildings, to be used for the selection of design targets. Other recent examples of the use of quality classifications are provided by [Lillicrap et al. 2005] and [Boerstra et al. 2005]. The first paper describes progress on methodologies for certifying the energy performance of buildings in accordance with the European Energy performance of Buildings Directive. The latter presents a new Dutch adaptive thermal comfort guideline. The design oriented *objectives* to cover the problem II (the use of the simulation environment for design) are:

    (i)   the development and evaluation of (new) applications for performance-based design.

    (ii)   evaluation of the usability of this work in relation with integral building assessment as proposed by [Hendriks & Hens 2000] and [Hendriks et al. 2003].

    (iii)   providing a preliminary guideline for design-oriented users.

The *methodology* was to perform case studies based on actual performance based design problems. This resulted in four applications in total. Furthermore, the order of applications was based on an increase of complexity of the building systems and operation, i.e. the first application contained relative simple systems and the last (fourth) application had the most complex systems and operations.

Figure 1.1 provides an illustration of the methodology. The implementation of the three model categories into the simulation environment is visualised at the upper part of the figure. The case studies are visualised at the lower part of the figure as applications subtracted from the simulation environment.

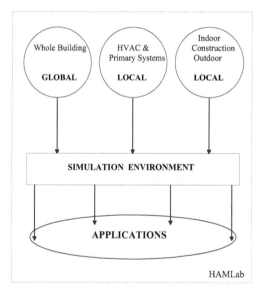

Figure 1.1 An illustration of the methodology

(This illustration is throughout the thesis used as background for several figures)

1.5 OUTLINE OF THE THESIS

Figure 1.2 shows a schematic overview of the thesis

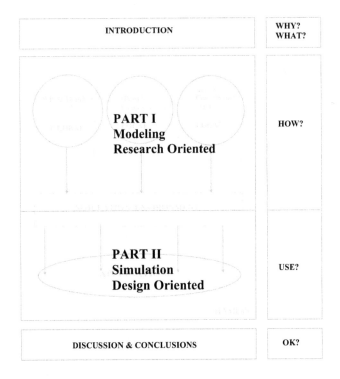

Figure 1.2 A schematic overview of the thesis

Part I presents the research oriented modeling volume, focussing on *how* the simulation environment can contribute to solve the previously mentioned common modeling problems.

Part II presents the case studies, focussing on the *use* of the simulation environment for performance based design.

The outline of the *Chapters 2 through 8* is presented at the prefaces of Part I and II. These chapters contain original papers already published in peer reviewed journals or on international conferences. In order to streamline the papers/chapters for the thesis, the following minor modifications are made: First, obsolete names are replaced i.e. WaVo by HAMBase; FemLab by ComSol. Second, the notations of the nomenclature and the references are unified. Third, some spelling errors, which were still present, are corrected. Fourth, the contents of some sections or paragraphs, which are already discussed, are replaced by references to previous sections. Fifth, footnotes are included for additional recent remarks regarding the content.

Chapter 9 revisits the problem statement and objectives and provides discussion and conclusions.

The *Literature* provides a comprehensive list of related work. The *Index of models* presents an overview of the more than 25 heat, air and moisture models included in this work.

The *Appendices* provide the status of several ongoing projects to be published in the near future.

REFERENCES

Ashino, R., Nagase, M., & Vaillancourt, R., 2000, Behind and beyond the MATLAB ODE Suite, *Report CRM-2651*

ASHRAE, 2005, *ASHRAE Handbook Fundamentals*, ISBN 1931862710

ASHRAE, 2003, *ASHRAE Handbook Applications* , ISBN 1931862222

Augenbroe, G., 2002, Trends in building simulation *Building and Environment 37,* pp891 – 902

Bartak, M., Baeusoleil, I., Clarke, J.A., Denev, J., Drkal, F., Lain, M., Macdonald, I.A., Melikov, A., Popiolek, Z. & Stankov, 2002, Integrating CFD and building simulation, *Building and Environment 37*, pp865 – 871

Boerstra, A.C., Raue, A.K., Kurvers, S.R., Linden, A.C., van der Hogeling, J.J.N.M. & de Dear, R.J., 2005, A new Dutch adaptive comfort guideline, *Conference on the Energy performance Brussels 21-23 September 2005*, 6p

Bradley, D. & Kummert, M., 2005, New evolutions in TRNSYS – a selection of version 16 features, *9TH IBPSA Conference Montreal*, pp107-113

Casas, W., Prölβ, K., & Schmitz, G., 2005, Modeling of desiccant assisted air conditioning systems *4^{tH} Modelica Conference. Hamburg March 7-8 2005* pp487-496

Citherlet, S., Clarke., J. A. & Hand, J., 2001, Integration in building physics simulation, *Energy and Buildings 33*, pp 451-461

Clarke, J.A., 2001, Domain integration in building simulation, *Energy and Buildings 33*, pp303-308

Crawley, D.B., Hand, J.W. Kummert, M. & Griffith B., 2005, Contrasting the capabilities of building performance simulation programs, *9TH IBPSA Conference Montreal* pp 231- 238 & *Report VERSION 1.0 July 2005*

Djunaedy, E., 2005, External coupling between building energy simulation and computational fluid dynamics, *PhD Thesis*, Eindhoven University of Technology

Ellis, M.W. & Mathews, E.H., 2002, Needs and trends in building and HVAC system design tools, *Building and Environment 37,* pp461 – 470

Felgner, F., Agustina, S., Cladera Bohigas, R., Merz, R., & Litz, L., 2002, Simulation of thermal building behaviour in Modelica, *2^{tH} Modelica Conference,* pp147-154

Felsmann, C., Knabe, G. & Werdin, H., 2003, Simulation of domestic heating systems by integration of trnsys in a matlab/simulink model, *6TH Conference System Simulation in Buildings Liege*, pp79-96

Felsmann, C. & Knabe, G., 2001. Simulation and optimal control strategie of HVAC systems, *7TH IBPSA Conference Rio de Janeiro*, pp1233-1239

Garcia-Sanz, M., 1997, A reduced model of central heating systems as a realistic scenario for analyzing control strategies, *Appl. Math modelling 21*, pp535-545

Gouda, M.M., Underwood, C.P. & Danaher, S., 2003, Modeling the robustness properties of HVAC plant under feedback control, *6TH Conference System Simulation in Buildings Liege*, pp511-524

Gough, M.A. 1999, A review of new techniques in building energy and environmental modeling, *BRE Report No. BREA-42, April 1999*

Hendriks, L. & Linden, K. van der, 2003, Building envelopes are part of a whole: reconsidering tradional approaches, *Building and environment 38*, pp 309-318.

Hendriks L., Hens H., 2000, Building Envelopes in a Holistic Perspective, IEA-Annex 32, Uitgeverij ACCO, *ISBN 90-75-741-05-7*

Hensen, J., Djunaedy, E., Radosevic, M. & Yahiaoui, A., 2004, Building performance for better design: some issues and solutions, *21TH PLEA Conference Eindhoven*, pp1185-1190

Hong T., Chou, S.K. & bong, T.Y., 2000, Building simulation: an overview of developments and information sources, *Building and Environment 35*, pp 347-361

IEA Annex 41, 2006, http://www.ecbcs.org/annexes/annex41.htm

Khoury El, Z., Riederer P., Couillland N., Simon J. & Raguin M., 2005, A multizone building model for Matlab/SimuLink environment, *9TH IBPSA Conference Montreal*, pp525-532

Liao, Z. & Dexter, A.L., 2004, A simplified physical model for estmating the average air temperature in multi-zone heating systems, *Building and environment 39*, pp 1013-1022.

Lillicrap D.C. & Davidson P.J., 2005, Building energy standards – tool for certification (bestcert) – pilot methodologies investigated. *Conference on the Energy performance Brussels 21-23 September 2005*, 6p

Mechaqrane, A. & Zouak, M., 2004, A comparison of linear and neutrl network ARX models applied to a prediction of the indoor temperature of a building. *Neural Computing & Applicication 13*, pp32-37.

Mendes, N., Oliveira, R.C.L.F., Araujo, H.X. & Coelho, L.S., 2003, A matlab-based simulation tool for building thermal performance analysis, *8TH IBPSA Conference Eindhoven*, pp855-862

Novak, P.R., Mendes, N. & N. Oliveira G.H.C, 2005, Simulation of HVAC plants in 2 brazilian cities using Matlab/SimuLink, *9TH IBPSA Conference Montreal*, pp859-866

Nytsch-Geusen, C.N., Nouidui, T. Holm, A. & Haupt W., 2005, A hygrothermal building model based on the object-orented modeling language modelica. *9TH IBPSA Conference Montreal*, pp867-873.

Oberkampf, W.L. & Trucano, T.G., 2002, Verification and validation in computational fluid dynamics, *Progress in Areospace Sciences 38*, pp209-272

Olsson, H., 2005, External Interface to Modelica in Dymola, *4^{tH} Modelica Conference. Hamburg,* pp603-611

Pohl, S.E. & Ungethum, J., 2005, A simulation management environement for Dymola. *4^{tH} Modelica Conference. Hamburg,* pp. 173-176.

Rabl, A., 1988, Parameter estimation in buildings: Methods for dynamic analysis of measured energy use, *Journal of Solar Energy Engineering 110*, pp52-66

Riederer P., 2005, MatLab/SimuLink for building and HVAC simulation – state of art. *9TH IBPSA Conference Montreal*, pp1019-1026

Sahlin, P., Eriksson, L., Grozman, P., Johnsson, H., Shapovalov, A. & Vuolle, M., 2004, Whole-building simulation with symbolic DAE equations and general purpose solvers, *Building and Environment 39*, pp949-958

Saldamli, L., Bachmann, B., Fritzson, P. & Wiesmann, H. 2005. A framework for describing and solving PDE Models in Modelica, *4^{tH} Modelica Conference. Hamburg,* pp113-122.

Sasic Kalagasidis, A., 2004, HAM-Tools, An Integrated Simulation Tool for Heat, Air and Moisture Transfer Analyses in Building Physics, *PhD Thesis*, ISSN 1400-2728

Schijndel, A.W.M. van & Wit, M.H. de, 1999, A building physics toolbox in MatLab, *7TH Symposium on Building Physics in the Nordic Countries Goteborg*, pp81-88

Schijndel, A.W.M. van, 1997, Building physics applications in Matlab, *1ST Benelux MatLab Usersconference Amsterdam*, Chapter 11

Schwab, M. & Simonson, C., 2004, Review of building energy simulation tools that include moisture storage in building materials and HVAC systems, Draft Report IEA Annex 41, Zurich

U.S. Department of energy 2006
http://www.eere.energy.gov/buildings/tools_directory/, visited February 1, 2006

Virk G.S., Azzi, D., Azad, A.K.M. & Loveday D.L., 1998, Recursive models for multi-roomed bms applications. *UKACC Conference on Control 1-4 september 1998*, pp1682-1687

Weitzmann, P. Sasic Kalagasidis, A., Rammer Nielsen, T. Peuhkuri, R. & Hagentoft C-E., 2003, Presentation of the international building physics toolbox for SimuLink, *8 TH IPBSA Conference Eindhoven*, pp1369-1376

Wit M.H. de & Driessen, H.H., 1988, ELAN A Computer Model for Building Energy Design. *Building and Environment 23*, pp.285-289

Wit, M.H. de, 2006, HAMBase, Heat, Air and Moisture Model for Building and Systems Evaluation, *Bouwstenen 100*, ISBN 90-6814-601-7 Eindhoven University of Technology

Yahiaoui, A., Hensen, J. Soethout, L. & Paassen, D, 2005, Interfacing building performance simulation with control modeling using internet sockets, *9TH IBPSA Conference Montreal 2005*, pp1377-1384

Yu, B., 2003, Level-Oriented Diagnosis for indoor Climate Installations, *PhD thesis*, ISBN 90-9017472-9

Zhai, Z., Chen, Q., Haves, P. & Klems, H., 2002, On approaches to couple energy simulation and computational fluid dynamics programs, *Building and Environment 37*, pp 857 – 864

Zhai Z. & Chen, Q., 2003, Solution of iterative coupling between energy simulation and CFD programs, *Energy and Buildings 35*, pp 493-505

Zhai Z. & Chen, Q., 2005, Performance of coupled building energy and CFD, *Energy and Buildings 37*, pp 333-344

SIMULATION TOOLS

BSIM	http://www.bsim.dk
Dymola	http://www.dynasim.se
Energyplus	http://www.eere.energy.gov/ buildings/energyplus/cfm/reg_form.cfm
ESP-r	http://www.esru.strath.ac.uk/Programs/ESP-r.htm
Comsol	http://www.comsol.com/
FlexPDE	http://www.pdesolutions.com/
HAMLab	http://sts.bwk.tue.nl/hamlab/
IDA ICE	http://www.equa.se/ice/intro.html
MathCad	http://www.mathsoft.com/
Mathematica	http://www.wolfram.com/
Matlab	http://www.mathworks.com/
Maple	http://www.maplesoft.com/
Modelica	http://www.modelica.org/
Optimization Toolbox	http://www.mathworks.com
SIMBAD	http://software.cstb.fr/soft/present.asp?page_id=us!SIMBAD
SimuLink	http://www.mathworks.com/
TRNSYS	http://sel.me.wisc.edu/trnsys/

PART I. RESEARCH.

THE SIMULATION ENVIRONMENT AS

A SUBJECT OF AND TOOL FOR RESEARCH

Preface

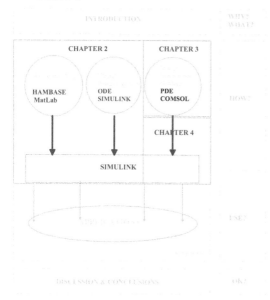

Part I presents the research oriented modeling volume, focussing on *how* the simulation environment can contribute to solve the previously mentioned common modeling problems.

The methodology to develop an integrated HAM modeling and simulation environment was to implement three components: whole building, HVAC/primary systems and airflow/construction models. This is presented in the next three chapters.

Chapter 2: covers whole building and HVAC/primary systems. The whole building model is implemented as follows: starting point is the already existing whole building model HAMBase in MatLab. In order to integrate the HAMBase model into SimuLink, the discrete HAMBase model is transformed into a continuous model. This continuous model and also the HVAC and primary systems models are mathematical modelled by ODEs, which are implemented into SimuLink by the use of S-Functions.

The third component (airflow and construction models) is presented in the next two chapters.

Chapter 3 presents the modeling of indoor airflows and hygrothermal construction responses by PDEs in Comsol. This chapter presents how these PDE based models can be implemented and simulated using Comsol.

Chapter 4 provides the integration of PDE based models for airflow and hygrothermal construction responses into SimuLink, including Comsol models of convective airflow and thermal bridges integrated into controller models of SimuLink.

Chapter 2

Advanced simulation of building systems and control with SimuLink

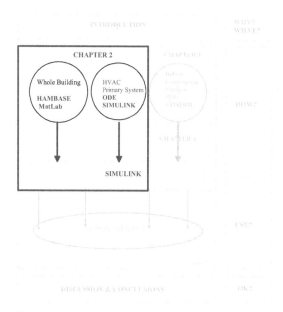

This chapter covers the implementation of the whole building model HAMBase and HVAC and primary systems models into SimuLink by the advanced use of S-Functions, facilitated by the Matlab/SimuLink environment. An existing indoor climate model is implemented in an S-Function, consisting of a continuous part with a variable time step and a discrete part with a fixed time step. The heating systems, including a heat pump, an energy roof and thermal energy storage (TES) are modelled as continuous systems using S-Functions. All presented models are validated. The advantages of S-Functions are evaluated. It demonstrates the powerful and flexible use of MatLab/SimuLink for simulating building and systems models

(A.W.M. van Schijndel & M.H. de Wit, published in proceedings of the 8 [TH] International IBPSA Conference, 2003, Vol. 3 pp. 1185- 1192)

2.1 INTRODUCTION

SimuLink [Mathworks 1997] is a software package for modeling, simulating, and analyzing dynamical systems. It supports linear and non-linear systems, modelled in continuous time, sampled time or a hybrid of the two. SimuLink includes a block library of sinks, sources, linear and non-linear components and connectors. Each block within a SimuLink model has the following general characteristics: a vector of inputs, **i**, a vector of outputs, **o**, and a vector of states **x**, as shown by Figure 2.1.

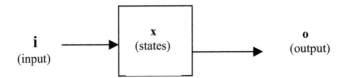

Figure 2.1 SimuLink block definition.

SimuLink facilitates hierarchical top-down and bottom-up modeling approaches. By double-click on blocks the level of model detail is increased. However, if a SimuLink model has a lot of blocks and a lot of levels, the model organization and how parts interact can become quite difficult to understand. S-functions (system functions) provide another way to create SimuLink models. Algorithms in MatLab or C can be implemented in S-functions. The main advantage of using S-functions is that users can build general-purpose blocks that can be used many times in a model, varying parameters with each instance of the block. SimuLink makes repeated calls during specific stages of simulation to each routine in the model, directing it to perform tasks such as computing its outputs, updating its discrete states, or computing its derivatives. Table 2.I shows these stages.

Table 2.I. Stages in S-functions [Mathworks 1997]: C = Continuous, D = Discrete, O = Output, t = time).

Simulation stage	S-Function Routine	Flag	Computation
Initialization	mdlInitializeSizes	flag = 0	$x = x_0$
Calculation of outputs	mdlOutputs	flag = 3	$x = x_C + x_D$, $y = f_O(t,x,u)$
Update discrete states	mdlUpdate	flag = 2	$x_D = f_D(t,x,u)$
Calculation of derivatives	mdlDerivatives	flag = 1	$dx_C/dt = f_C(t,x,u)$

The use of S-Functions is demonstrated by the implementation of buildings systems and the building model HAMBase (Heat, Air and Moisture model for Building and Systems Evaluation) in SimuLink. The main objective of HAMBase is the simulation of the thermal and hygric indoor climate. A brief description of the model is given in Section 2.2. A more detailed description can be found in [de Wit 2006]. The application of S-Functions in SimuLink is shown in Section 2.3 and Section 2.4. In Section 2.3, a hybrid (continuous/discrete) building zone model based on HAMBase is presented. Section 2.4 presents a case study of an energy roof assisted heating system controlled with an on/off controller. The system includes continuous models of an energy roof, a heat pump and a thermal energy storage system (TES). In Section 2.5 simulation results of the presented models are confronted with measurements. The advantages of S-Functions are evaluated at the Conclusions. Finally, the complete S-function code for the presented heat pump model is given in the appendix.

2.2 THE BUILDING MODEL HAMBASE

The physics of the HAMBase model is based on ELAN, a computer model for building energy design. [Wit et al. 1988]. More recently, the ELAN model, together with an analog hygric model, has been implemented in MatLab. A short summary of the HAMBase model is now presented, further details can be found in [de Wit 2006].

The HAMBase model uses an integrated sphere approach. It reduces the radiant temperatures to only one node. This has the advantage that also complicated geometries can easily be modelled. In figure 2.2, the thermal network is shown.

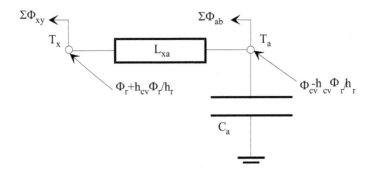

Figure 2.2 The room model as a thermal network

T_a is the air temperature and T_x is a combination of air and radiant temperature. T_x is needed to calculate transmission heat losses with a combined surface coefficient. h_r and h_{cv} are the surface weighted mean surface heat transfer coefficients for convection and radiation. Φ_r and Φ_{cv} are respectively the radiant and convective part of the total heat input consisting of heating or cooling, casual gains and solar gains.[*]

For each heat source a convection factor can be given. For air heating the factor is 1 and for radiators 0.5. The factor for solar radiation depends on the window system and the amount of radiation falling on furniture. C_a is the heat capacity of the air. L_{xa} is a coupling coefficient [Wit et al. 1988]:

$$L_{xa} = A_t h_{cv} \left(1 + \frac{h_{cv}}{h_r}\right) \tag{2.1}$$

$\sum \Phi_{ab}$ is the heat loss by air entering the zone with an air temperature T_b. A_t is the total area. In case of ventilation T_b is the outdoor air temperature. $\sum \Phi_{xy}$ is transmission heat loss through the envelope part y. For external envelope parts T_y is the sol-air temperature for the particular construction including the effect of atmospheric radiation.

---

[*]Please note that the model presented in figure 2.2 is a result of a delta-star transformation. More details are provided in Appendix E

The thermal properties of the wall and the surface coefficients are considered as constants, so the system of equations is linear. For this system the heat flow entering the room can be seen as a superposition of two heat flows: one resulting from T_y with $T_x=0$ and one from T_x with $T_y=0$. The next equations are valid in the frequency domain:

$$\Phi_x = -\Phi_{xx} + \Phi_{yx} = -Y_x T_x + Y_{xy}\left(T_y - T_x\right)$$
$$\Phi_y = \Phi_{yy} + \Phi_{yx} = Y_y T_y + Y_{xy}\left(T_y - T_x\right) \tag{2.2}$$

The heat flow (Φ_{yx}) caused by the temperature difference ΔT_{yx} is modelled with a fixed time step (1 hour) and response factors. For $t = tn$:

$$\Phi_{yx}(tn) = L_{yx}\Delta T_{yx}(t_n) + \Delta\Phi_{yx}(t_n)$$

$$\Delta\Phi_{yx}(t_n) = a_1\Delta T_{yx}(t_{n-1}) + a_2\,\Delta T_{yx}(t_{n-2}) + b_1\Delta\Phi_{yx}(t_{n-1}) + b_2\Delta\Phi_{yx}(t_{n-2}) \tag{2.3}$$

The next equation for the U-value of the wall is valid:

$$A_{xy}U_{xy} = L_{xy}+ (a_1+ a_2)/(1-b_1-b_2) \tag{2.4}$$

For glazing, thermal mass is neglected:

$$L_{yx}= A_{glazing}U_{glazing} \quad (a_1=a_2=b_1=b_2=0) \tag{2.5}$$

The heat flow at the inside of a heavy construction is steadier than in a lightweight construction. In such case L_{yx} will be close to zero. In the model L_{yx} is a conductance (so continuous) and Φyx are discrete values to be calculated from previous time steps. For adiabatic envelope parts $\Phi_{yx} = 0$. In the frequency domain, the heat flow Φ_{xx} from all the envelope parts of a room can be added:

$$\Phi_{xx}(tot) = -T_x \sum Y_x \tag{2.6}$$

The admittance for a particular frequency can be represented by a network of a thermal resistance $(1/L_x)$ and capacitance (C_x) because the phase shift of Y_x can never be larger than $\pi/2$. To cover the relevant set of frequencies (period 1 to 24 hours) two parallel branches of such a network are used giving the correct admittance's for cyclic variations with a period of 24 hours and of 1 hour. This means that the heat flow $\Phi_{xx}(tot)$ is modelled with a second order differential equation. For air from outside the room with temperature T_b a loss coefficient L_v is introduced. The model is summarized in figure 2.3

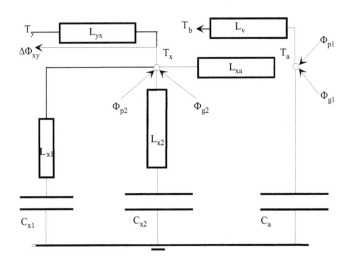

Figure 2.3 The thermal model for one zone

In a similar way a model for the air humidity is made. Only vapour transport is modelled, the hygroscopic curve is linearized between RH 20% and 80%. The vapour permeability is assumed to be constant. The main differences are: a) there is only one room node (the vapour pressure) and b) the moisture storage in walls and furniture, carpets etc is dependent on the relative humidity and temperature.

The HAMBase model has been subjected to the ASHRAE test [ASHRAE 2001] with satisfactory results. For further details, see Table 2.II.

Table 2.II Comparison of the room model with some cases of the standard test [ASHRAE 2001][*]:

Case Nr.	Simulation of	model result	min.test	max. test
600ff	mean indoor temperature [°C]	24.8	24.2	25.9
600ff	minimum indoor temperature [°C]	-19.1	-18.8	-15.6
600ff	maximum indoor temperature [°C]	64.7	64.9	69.5
900ff	mean indoor temperature [°C]	24.8	24.5	25.9
900ff	minimum indoor temperature [°C]	-5.5	-6.4	-1.6
900ff	maximum indoor temperature [°C]	43.1	41.8	44.8
600	annual cooling [MWh]	6.7	6.1	8.0
600	annual heating [MWh]	5.4	4.3	5.7
600	peak heating [kW]	4.1	3.4	4.4
600	peak cooling [kW]	6.3	6.0	6.6
900	annual heating [MWh]	1.9	1.2	2.0
900	annual cooling [MWh]	2.6	2.1	3.4
900	peak cooling [kW]	3.7	2.9	3.9

2.3 THE HAMBASE MODEL IN SIMULINK

A major recent improvement is the development of a HAMBase model in SimuLink. The model consists of a continuous part with a variable time step and a discrete part with a time step of one hour. For the HVAC installation and the room response on indoor climatic variations a continuous model is used (see figure 2.3). For the external climate variations a discrete model is used. The main advantages of this numeric hybrid approach are:

> a) The dynamics of the building systems where small time scales play an important role (for example on/off switching) are accurately simulated.

> b) The model becomes time efficient as the discrete part uses 1-hour time steps. A yearly based simulation takes 2 minutes on a Pentium III, 500 MHz computer.

> c) The moisture (vapour) transport model of HAMBase is also included. With this feature, the (de-) humidification of HVAC systems can be simulated.

---

[*] Cases 600 & 900 contain repectively lightweight and heavy constructions; ff means free floating

The heat transport part of HAMBase SimuLink is also validated by the ASHRAE test [ASHRAE 2001]. The results of the HAMBase SimuLink model are identical to the results of the HAMBase model. The moisture transport part is not yet completely validated. Preliminary results [see Section 7.3.1] show a good agreement between model and measurement.

In figure 2.4, an example of the use of a 1-zone HAMBase SimuLink model is demonstrated.

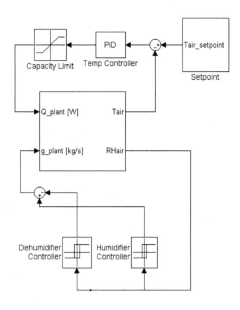

Figure 2.4 The HAMBase SimuLink model, including the following controllers: a PID Tair-controller with limited heating/cooling and on/off RHair-controllers with (de) humidification.

The inputs of the HAMBase SimuLink model are heat flow and moisture flow and the outputs are air temperature and relative humidity. In this example the air temperature is controlled by a limited PI controller and the relative humidity bounds are controlled by on/off controllers.

2.4 THE HEATING SYSTEM IN SIMULINK

A short introduction of the heating system is given now. An energy roof collector is cooled by a heat pump so that its surface temperature will often be below the ambient air temperature. This has the advantage that besides solar energy, also energy is gained from the ambient air. In the Netherlands the winters are mild and humid with little sunshine, so the system may be promising. In the past, several configurations of an energy roof with a focus on the convective heat recovery from ambient air have been investigated at the GEO test site of the University [Jong et al. 2000]. The thermal energy storage (TES) is located at the cold side of the heat pump so instead of heat loss even heat gain is possible. In figure 2.5, an outline is given of the energy roof system.

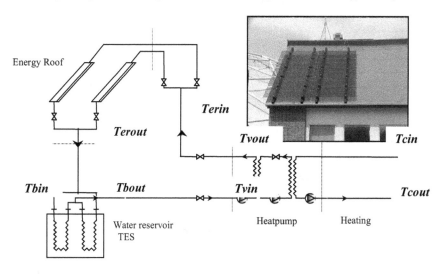

Figure 2.5 The energy roof system, including water temperatures (bold italic).

The system has two identical roof surfaces, one facing south and one facing north. This enables to investigate whether an energy roof facing north, without direct solar radiation is cost effective. Circulating the cooling fluid through the energy roof charges the TES. Discharging the TES is accomplished by passing the cooling fluid through the heat pump. (In the real system charging and discharging can be done at the same time, when

27

the energy roof is extracting heat from outside and there is a simultaneous heat demand from the dwelling). The collector of the test site consists of a simple perforated plate designed primarily for convective heat transfer.

The measurements of [Jong et al. 2000] are used for the determination of the constants in the component models. The models[*] are:

Heat pump model:

$$
\left[
\begin{array}{l}
COP = k \cdot \dfrac{0.5 \cdot T_{cin} + 0.5 \cdot T_{cout} + 273.15}{(0.5 \cdot T_{cin} + 0.5 \cdot T_{cout}) - (0.5 \cdot T_{vin} + 0.5 \cdot T_{vout})} \\[2ex]
C_c \dfrac{dT_{cout}}{dt} = MF_{cin} \cdot c_w \cdot (T_{cin} - T_{cout}) + COP \cdot E_{hp} \\[2ex]
C_v \dfrac{dT_{vout}}{dt} = MF_{vin} \cdot c_w \cdot (T_{vin} - T_{vout}) - (COP - 1) \cdot E_{hp}
\end{array}
\right.
\tag{2.7}
$$

Where T is temperature [°C], COP Coefficient of Performance [-], k heat pump efficiency determined from the measurements at the GEO test site, (k=0.4), c_w specific heat capacity of water, C heat capacity of the water and pipes in the heat pump ($Cv=Cc$ 10^5 [J/K]), t time[s], MF mass flow [kg/s], E_{hp} heat pump electric power supply (1200 W). Subscript c means water at the condenser, v water at the evaporator, in, incoming, out, outgoing. The complete S-function code for the heat pump model is given in the appendix.

Energy roof model:

$$
C_{er} \dfrac{dT_{erout}}{dt} = MF_{erin} \cdot c_w \cdot (T_{erin} - T_{erout}) + k_1 \cdot E_{solar} - k_2 \cdot (\dfrac{T_{erin} + T_{erout}}{2} - T_e)
\tag{2.8}
$$

Where E_{solar} is solar irradiance [W/m²], k_1 and k_2 empirical determined parameters (k_1=0.8 m2 and k_2=125 W/K). Subscript er means water at energy roof, e exterior.

---

[*] The models presume perfect mixing and no heat losses

Thermal energy storage:

$$m \cdot c_w \cdot \frac{dT_{bout}}{dt} = MF_{bin} \cdot c_w \cdot T_{bin} - MF_{bout} \cdot c_w \cdot T_{bout} \qquad (2.9)$$

Where *m* is the mass of storage [kg], Subscript b means water in TES.

The models (2.7), (2.8) and (2.9) are implemented in SimuLink also using S-functions.

A complete example of the S-Function of the heat pump model can be found in the appendix.

2.5 ANALYSIS

With the parameters found from the measurements the calculated performances of the components are compared with measurements. The input for the models are: the measured incoming and outgoing mass flows, incoming water temperatures (and for the energy roof also the external temperature and the irradiance on the inclined surface). Figure 2.6-2.8 show that the models of the components predict the outgoing water temperatures well:

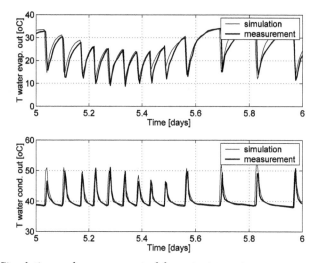

Figure 2.6 Simulation and measurement of the outgoing water temperatures of the heat pump

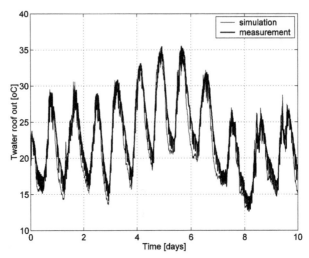

Figure 2.7 Simulation and measurement of the outgoing water temperatures of the energy roof

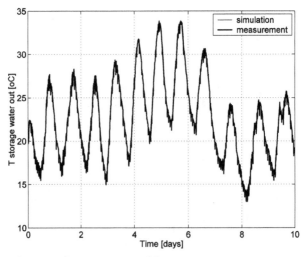

Figure 2.8 Simulation and measurement of the outgoing water temperatures of the TES

In figure 2.9, the complete energy roof system, connected to the building zone model in SimuLink, is presented.

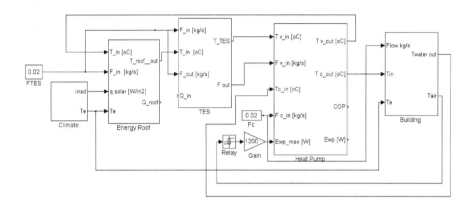

Figure 2.9 The complete energy roof system in SimuLink.

The control strategy is simple. All mass flows (water) are kept constant (0.02 kg/s) and the heat pump is controlled by the internal temperature and an on/off switch (Relay). The external temperature and the irradiance on the inclined surface are used as input for the simulation of the complete system.

In figure 2.10, the simulated indoor air temperature using the model of figure 2.9, is compared with measurements.

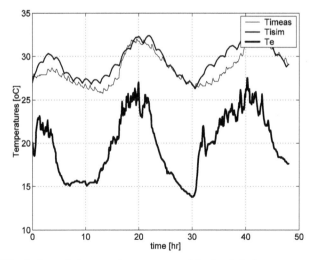

Figure 2.10 The indoor air temperature, measured & simulated

31

There are differences between the calculated values and the measured ones. This is probably due to a different control of the mass flows between the model and the reality, where the mass flows are also switched on and off. This has not been implemented yet and is left over for future research[*]. Figure 2.11, shows the temperatures for a 48 hours period. These include: Incoming and outgoing water temperatures of the evaporator and condenser, outgoing water temperature of the energy roof and internal and external air temperatures.

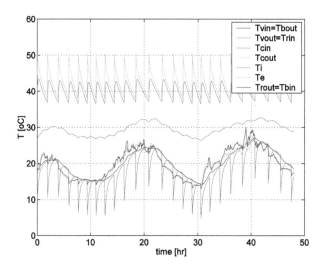

Figure 2.11 The temperatures for a 48 hours period.

The model of figure 2.9, has successfully been used for optimizing the energy roof system [Blezer 2003]. Up to now, it may be concluded that cost-efficient application of a heat pump in a dwelling is best achieved by bivalent systems. The capacity of the heat pump is limited then to about 30% of the total required maximal heating capacity.

---

[*] More details can be found in appendix B1 & D1

2.6 CONCLUSIONS

The following applications of S-functions in SimuLink for building systems component simulation have been evaluated:

* A hybrid (continuous/discrete) building zone model capable of simulating the thermal and hygric indoor climate. The main advantages of this model are: a) the dynamics of the building systems of time scales less than an hour are accurately simulated, b) the model becomes time efficient. The building zone model is validated with the ASHRAE test [ASHRAE 2001].

* Continuous models of a heat pump, an energy roof and a TES (Thermal Energy Storage). The main advantages of this approach are: a) a clear relation between mathematical model (system of Ordinary Differential Equations (ODEs)) and computer code in the S-functions, b) the state of art ODE solvers of MatLab gives accurate solutions. The models are compared with measurements are acceptable.

* A complete energy roof and building model containing all the above mentioned components with a preliminary simple control strategy. Future models will include more advanced control strategies in order to get more realistic simulation results and to validate the complete model.

The evaluation illustrates the powerful and flexible nature of Matlab/SimuLink for simulating building systems models.

REFERENCES

ASHRAE, 2001, Standard method of test for the evaluation of building energy analysis computer programs, standard 140-2001.

Blezer, I., 2002, Modeling, Simulation and optimization of a heat pump assisted energy roof system. Master thesis (in Dutch), Univ. of Tech. Eindhoven, group FAGO.

Jong, J. de, A.W.M. van Schijndel & C.E.E. Pernot, 2000, Evaluation of a low temperature energy roof and heat pump combination, Int. Building Physics Conf. Eindhoven, 18-21 Sept. 2000, pp99-106

Mathworks Inc., 1997, SimuLink version 2. Reference Guide

Wit M.H. de and H.H. Driessen, 1988, ELAN A Computer Model for Building Energy
Design. Building and Environment, Vol.23, No 4, pp.285-289

Wit, M.H. de, 2006, HAMBase, Heat, Air and Moisture Model for Building and
Systems Evaluation, Bouwstenen 100, ISBN 90-6814-601-7 Eindhoven
University of Technology

APPENDIX

A complete example how to model a system of ODEs with an S-function of SimuLink
is shown for the heat pump model. The first step is to define the input-output definition
of the model. This is presented in Table 2.III.

Table 2. III. The input-output definition of the heat pump model

Variable name	Input (u) / Output (y)	Description
T_{vin}	u(1)	Incoming water temperature at the evaporator [^{o}C]
MF_{vin}	u(2)	Incoming mass flow at the evaporator [kg/s]
T_{cin}	u(3)	Incoming water temperature at the condenser [^{o}C]
MF_{cin}	u(4)	Incoming mass flow at the condenser [kg/s]
Ehp	u(5)	Power of electrical supply [W]
k	u(6)	efficiency [-]
T_{vout}	y(1)	Outgoing water temperature at the evaporator [^{o}C]
T_{cout}	y(2)	Outgoing water temperature at the condenser [^{o}C]
COP	y(3)	Coefficient Of Performance [-]

The second step is to formulate a mathematical model by a system of ODEs. This is
done using (2.7). The third step is to implement the mathematical model into a (S)ystem
function, a programmatic description of a dynamic system. Details about this subject
can be found in [Mathworks 1997]. Figure 2.12 shows the final SimuLink model.
Figure 2.13 shows the program code.

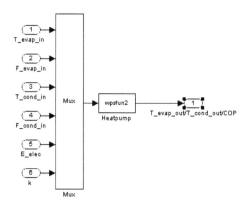

Figure 2.12 The SimuLink model of the heat pump.

```
function [sys,x0,str,ts] = wpsfun2(t,x,u,flag)
%u(1)=Tvin [oC], u(2)=MFvin [kg/s], u(3)=Tcin [oC],
%u(4)=MFcin [kg/s], u(5)=Ehp [W], u(6)=k [-]
%y(1)=Tvout [oC] (=x(1)), y(2)=Tcout [oC] (=x(2)) %y(3)=COP
    switch flag,
        case 0,        [sys,x0,str,ts]=mdlInitializeSizes;
        case 1,        sys=mdlDerivatives(t,x,u);
        case 3,        sys=mdlOutputs(t,x,u);
        case { 2, 4, 9 },    sys=[];
    end
function [sys,x0,str,ts]=mdlInitializeSizes
    sizes.NumContStates  = 2; % Number of Cont. states
    sizes.NumDiscStates  = 0; % Number of Disc. states
    sizes.NumOutputs     = 3; % Number of Outputs
    sizes.NumInputs      = 6; % Number of Inputs
    x0  = [10; 10];                   % Initial values
function sys=mdlDerivatives(t,x,u)
        Tvm=(u(1)+x(1))/2;    Tcm=(u(3)+x(2))/2;
        COP=u(6)*(273.15+Tcm)/(Tcm-Tvm);
        Cc=200000;Cv=200000;cv=4200;cc=4200;
        xdot(1)=(1/Cv)*(u(2)*cv*(u(1)-x(1))-(COP-1)*u(5));
        xdot(2)=(1/Cc)*(u(4)*cc*(u(3)-x(2))+COP*u(5));
    sys = [xdot(1); xdot(2)];
function sys=mdlOutputs(t,x,u)
        Tvm=(u(1)+x(1))/2;    Tcm=(u(3)+x(2))/2;
        COP=u(6)*(273.15+Tcm)/(Tcm-Tvm);
    sys = [x(1); x(2); COP];
```

Figure 2.13 The code of the heat pump model used at the S function.

Chapter 3

Modeling and solving building physics problems with Comsol

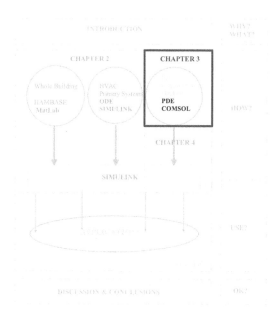

This Chapter covers the modeling of indoor air and constructions by PDEs in Comsol. This software is designed to simulate systems of coupled PDEs which may be 1D, 2D or 3D, non-linear and time dependent. An important feature of Comsol is that the user can focus on the model (PDE coefficients on the domain and boundary) and does not have to spend much time on solving and visualization. In this chapter, 4 cases are considered. First, in order to illustrate how Comsol works, an example including the complete code for solving as well as the results are given for a simple 2D steady state heat transfer problem. In the next 2 cases, the reliability is tested for two very different building physics problems: A 2D dynamic airflow problem, modelled using Navier Stokes and buoyancy equations, and a 1D dynamic non-linear moisture transport in a porous material. These simulation results are validated and show a good agreement with measurements. In the last case, Comsol's capability of simulating 3D problems is shown by a dynamic combined heat and moisture transport problem. This example is a 3D extension of a given 2D problem from IEA Annex 24 [Kunzel 1996]. For all models the crucial part of the codes (geometry, PDEs and boundary specifications) are given. The Comsol software is written in the MatLab environment [Mathworks 1998] and therefore it is possible to use the visualization tools, toolboxes and all other programs written in MatLab. The evaluation illustrates the powerful and flexible nature of Comsol for solving scientific and engineering building physics problems.

(A.W.M. van Schijndel, published in Building and Environment, 2003, vol 38/2, pp319-327)

3.1 INTRODUCTION

Many scientific problems in building physics can be described by PDEs. There are a lot of software programs available in which one specific PDE is solved. They are developed in order to get the simulation results in a short time and most often a lot of effort has been put into the simplicity of input of data, e.g. geometrical data. A disadvantage is that they often are not very flexible when the user wants to change or combine models. Another drawback is that they most often act as black boxes. Another category of commercially available software like Comsol [Comsol 2006] is developed specifically for solving PDEs where the user in principle can simulate any system of coupled PDEs. Practical physics/engineering problems in the area of heat transfer, electromagnetism, structural mechanics and fluid dynamics can be solved with the software. The practical problems solved in this chapter are: a 2D thermal bridge problem, a 1D moisture transport problem, a 2D airflow problem and a 3D combined heat and moisture transport problem. One of the main advantages of Comsol is that the user can focus on the model (PDE coefficients on the domain and the domain boundary) and does not have to spend much time on solving and visualization. The scientist can concentrate on the physics behind the models and the engineer can calculate details for designing purposes using validated models.

Section 3.2 shows some features of Comsol and how it works by a simple 2D steady state heat transfer problem. In Section 3.3 the quality of the numerical solvers is tested by solving two very different building physics problems: A 2D dynamic airflow problem and a 1D dynamic non-linear moisture transport in a porous material. The solutions are compared with measurements. There are not many software packages available which are capable of simulating 3D dynamic combined heat and moisture transport problems. In Section 3.4 it is shown how this can be done using Comsol.

3.2 HOW COMSOL WORKS

Comsol [Comsol 2006] is a toolbox written in MatLab [Mathworks 1998]. It solves systems of coupled PDEs (up to 32 independent variables). The specified PDEs may be non-linear and time dependent and act on a 1D, 2D or 3D geometry. The PDEs and boundary values can be represented by two forms. The coefficient form is as follows[*]:

$$d_a \frac{\partial u}{\partial t} - \nabla \cdot (c\nabla u + \alpha u - \gamma) + \beta \nabla u + au = f \qquad in \ \Omega \qquad (3.1a)$$

$$\underline{n} \cdot (c\nabla u + \alpha u - \gamma) + qu = g - \lambda \qquad\qquad on \ \partial\Omega \quad (3.1b)$$

$$hu = r \qquad\qquad on \ \partial\Omega \quad (3.1c)$$

The first equation (3.1a) is satisfied inside the domain Ω and the second (3.1b) (generalized Neumann boundary) and third (3.1c) (Dirichlet boundary) equations are both satisfied on the boundary of the domain $\partial\Omega$. n is the outward unit normal and is calculated internally. λ is an unknown vector-valued function called the Lagrange multiplier. This multiplier is also calculated internally and will only be used in the case of mixed boundary conditions. The coefficients d_a , c, α , β, γ, a, f, g, q and r are scalars, vectors, matrices or tensors. Their components can be functions of the space, time and the solution u. For a stationary system in coefficient form $d_a = 0$. Often c is called the diffusion coefficient, α and ß are convection coefficients, a is the absorption coefficient and γ and f are source terms.

The second form of the PDEs and boundary conditions is the general form:

$$d_a \frac{\partial u}{\partial t} + \nabla \cdot \Gamma = F \qquad\qquad in \ \Omega \qquad (3.2a)$$

$$-\underline{n} \cdot \Gamma = G + \lambda \qquad\qquad on \quad \partial\Omega \qquad (3.2b)$$

$$R = 0 \qquad\qquad on \quad \partial\Omega \qquad (3.2c)$$

---

[*] The symbols provided by the Comsol modeling guides are also used here.

The coefficients Γ and F can be functions of space, time, the solution u and its gradient. The components of G and R can be functions of space, time, and the solution u.

Given the geometry, an initial finite element mesh is automatically generated by triangulation of the domain. The mesh is used for discretisation of the PDE problem and can be modified to improve accuracy. The geometry, PDEs and boundary conditions are defined by a set of fields similarly to the structure in the language C. A graphical user interface is used to simplify the input of these data. For solving purposes Comsol contains specific solvers (like static, dynamic, linear, non-linear solvers) for specific PDE problems.

Example: Comsol code and results of a 2D stationary thermal bridge

A 2D stationary thermal bridge problem is used as an example of how Comsol works. In figure 3.1 the geometry of the 2D thermal bridge problem is shown.

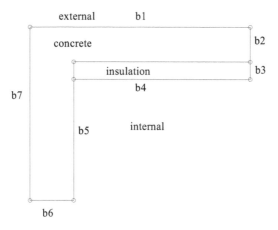

Figure 3.1 The geometry of the 2D thermal bridge example, see Table 3.1 for boundary conditions specifications.

In Table 3.I. the lengths and boundary conditions of each boundary segment are given.

Table 3.I. Boundary specifications of the 2D thermal bridge problem, where T is the temperature on the boundary and T_i and T_e respectively are the internal and external temperatures.

Boundary Segment	Boundary Type	Boundary Segment length [m]	Boundary condition [Wm^{-2}]
b1	external	1.0	$\varphi=hc_e*(T_e-T)$
b2	adiabatic	0.2	$\varphi=0$
b3	adiabatic	0.1	$\varphi=0$
b4	internal	0.8	$\varphi=hc_i*(T_i-T)$
b5	internal	0.7	$\varphi=hc_i*(T_i-T)$
b6	adiabatic	0.2	$\varphi=0$
b7	adiabatic	1.0	$\varphi=0$

The PDE model for the inside of the domain is given by:

$$\nabla \cdot (K\nabla T) = 0 \qquad (3.3)$$

Where K is thermal conductivity and T is temperature. Using the coefficient form (3.1a) and the model (3.3), it follows that u equals $\underline{T}$ and the coefficients of (3.1a) are all zero ($a= d_a = f= \alpha= \beta = \gamma =0$) except c. The c coefficient equals the heat conductivities at the sub domains concrete (K_{concr}) and insulation (K_{insul}). The boundary values are heat fluxes and so the Neumann condition is applied. For example, boundary condition b1: $f = h_{ce}*(T_e-T)$ is represented by taking $q = h_{ce}$, $g = h_{ce}*T_e$ in eq. (3.1b). Note that the term $\underline{n}\cdot c\blacktriangledown u$ in (3.1b) represents the heat flow into the domain and is calculated internally and the term λ in (3.1b) is zero because mixed boundary conditions are not applied in this example. Figure 3.2 shows the complete Comsol code.

```
        %CONSTANTS
hci=7.7;            %heat transfer coefficient internal
hce=25;            %heat transfer coefficient external
Ti=20;             %air temperature internal
Te=-10;            %air temperature external
Kconcr=1;          %heat conduction concrete
Kinsul=0.03;       %heat conduction insulation

        %GEOMETRY: poly2(XDATA,YDATA) ; 2D polygon
CONCR=poly2([0 0 1 1 0.2 0.2],[0 1 1 0.8 0.8 0]);       %concrete
INSUL=poly2([0.2 0.2 1 1],[0.7 0.8 0.8 0.7]);           %insulation
fem.geom=CONCR+INSUL;                                   %fem geometry
fem.dim=1;                                              %One component

        %COEFFICIENTS OF THE PDE/Boundary problem
fem.equ.c={Kconcr   Kinsul  };          % fem coefficient c
fem.bnd.g={hce*Te 0 0 hci*Ti hci*Ti 0 0};   % fem coefficient g
fem.bnd.q={hce  0    0 hci   hci   0 0};     % fem coefficient q
fem.mesh=meshinit(fem);                 % intialize mesh
fem.sol=femlin(fem);                    % solve, steady problem

        %OUTPUT MESH, SOLUTION
meshplot(fem)                           % plot mesh
q=posteval(fem,'u');                    % post processing data
postplot(fem,'tridata',q,'tribar','on')  % plot solution
```

Figure 3.2 The complete Comsol code for solving the 2D thermal bridge problem.

The default values of all PDE and boundary coefficients are 0. Also some comments (%) are included for better understanding of the code.

The initial mesh is presented in figure 3.3, and the solution[*] in figure 3.4

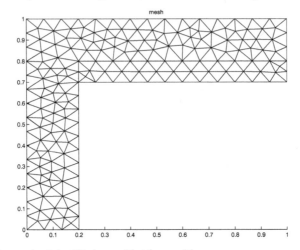

Figure 3.3 The mesh of the 2D thermal bridge problem.

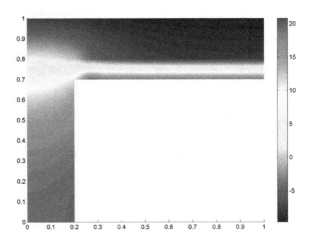

Figure 3.4 The solution (temperature distribution) of the 2D thermal bridge problem.

This example shows the transparency, easy-to-use and flexibility of PDE models in Comsol.

---

[*] Also temperature ratios and transmission coefficients can be calculated

3.3 TESTCASES FOR RELIABILITY

In [Comsol 2006] examples are already present which show the accuracy and reliability of Comsol. However there are no validations for typical building physics problems. Specific for validating building physics simulations in Comsol, this Section deals with two (very different) time dependent non-linear problems. Each problem is solved and compared with measured data.

3.3.1 1D Moisture transport in a porous material

The water absorption in an initially dry brick cylinder (length 24-mm) is studied. All sides except the bottom are sealed. This side is submerged in water at t=0 sec. The PDE and boundary conditions for this problem are shown in figure 3.5 (upper part). The corresponding Comsol code for defining the geometry, PDE and boundaries is shown in the lower part of figure 3.5

$$\frac{\partial \theta}{\partial t} = \nabla \cdot (D(\theta) \nabla \theta) \quad \text{on } 0 < x < 0.024,$$

$$\theta = 0.27 \quad \text{on } x = 0,$$

$$\frac{\partial \theta}{\partial x} = 0 \quad \text{on } x = 0.024$$

```
fem.geom=solid1([0 0.024]);        % 1D geometry
Dw=['(3.2e-9)*exp(29*u)'];         % diffusivity fired clay brick,type I
fem.dim=1;                         % one component
fem.equ.c={Dw};                    % c equals diffusivity
fem.equ.da={1};                    % da equals 1
fem.bnd.h={1  0};                  % boundary value
fem.bnd.r={0.27 0};                % boundary value
```

Figure 3.5 The PDE and boundary conditions (upper) and the corresponding part of the Comsol code (lower) of the 1D moisture transport problem.

The coefficient form (3.1) is used. The results of measured water absorption of several brick materials based on [Brocken 1998] are shown in figure 3.6, left hand side. The moisture profiles are shown versus lambda (lambda is defined by the position divided by the square root of the time). For each material the diffusivity as a function of the moisture content is given [Brocken 1998] and used to simulate the corresponding profiles.

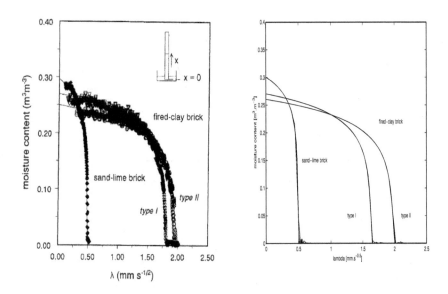

Figure 3.6 Measured and simulated moisture content profiles versus lambda of the 1D moisture transport problem. Left hand side: Measured moisture contents, Right hand side: Simulated moisture contents.

The simulation results in figure 3.6, right hand side, are compared with the measured profiles[*].

---

[*] Such a comparison is used to gain confidence in the model and simulation tool.

3.3.2 2D Airflow in a room

This example from [Sinha 2000] deals with the velocity and temperature distribution in a room heated by a warm air stream. In figure 3.7 the geometry and boundary conditions are presented.

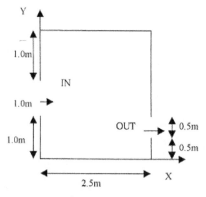

The boundary conditions are:
At the left, right, top and bottom walls:
u=0, v=0, T=0.
At the inlet:
u=1, v=0, T=1.
At the outlet :
Neuman conditions for u,v and T

Figure 3.7 The geometry and boundary conditions for the 2D-airflow problem.

The problem is modelled by 2D incompressible flow using the Boussinesq approximation with constant properties for the Reynolds and Grasshof numbers. The general form (3.2) is used for this type of non-linear problem. In figure 3.8, the PDE model, the corresponding PDE coefficients of (3.2a) and the corresponding part of the code are given.

U-momentum equation
$$\frac{\partial u}{\partial t} = -\frac{\partial(uu)}{\partial x} - \frac{\partial(vu)}{\partial y} - \frac{\partial p}{\partial x} + \frac{1}{Re}\nabla^2 u$$

V-momentum equation
$$\frac{\partial v}{\partial t} = -\frac{\partial(uv)}{\partial x} - \frac{\partial(vv)}{\partial y} - \frac{\partial p}{\partial y} + \frac{1}{Re}\nabla^2 v + \frac{Gr}{Re^2}T$$

Continuity equation
$$\frac{\partial u}{\partial x} + \frac{\partial v}{\partial y} = 0$$

Energy equation
$$\frac{\partial T}{\partial t} = -\frac{\partial(uT)}{\partial x} - \frac{\partial(vT)}{\partial y} + \frac{1}{Re\,Pr}\nabla^2 T$$

```
            %Variables:  u1=u; u2=v; u3=p; u4=T
            %       u1x = du1/dx, etc.
eta=1/Re;beta=Gr/(Re*Re);alpha=1/(Re*Pr); %parameters
fem.dim=4;
fem.equ.da={{1; 1; 0; 1}};
fem.equ.F={{    '-(u1.*u1x+u2.*u1y+u3x)';...
                '-(u1.*u2x+u2.*u2y+u3y)+beta*u4';...
                '-(u1x+u2y)';...
                '-(u1.*u4x+u2.*u4y)'}};
fem.equ.ga={{{      '-eta*u1x'; '-eta*u1y'};...
            {       '-eta*u2x'; '-eta*u2y'};...
                            0;...
            {       '-alpha*u4x'; '-alpha*u4y'}}};

fem.bnd.r={ {                   '-u1'; '-u2'; 0; '0-u4'} ...
            {                   '-u1'; '-u2'; 0; '0-u4'} ...
            {                   '1-u1'; '-u2'; 0; '1-u4'} ...
                    {           '-u1'; '-u2'; 0; '0-u4'} ...
                    {           '-u1'; '-u2'; 0; '0-u4'} ...
                    {           '-u1'; '-u2'; 0; '0-u4'} ...
                    {           0; 0; '-u3'; 0} ...
            {                   '-u1'; '-u2'; 0; '0-u4'} };

fem=femdiff(fem); % calculate divergence gamma on domain
```

Figure 3.8 The PDE model and the corresponding Comsol code for the 2D-airflow problem.

In [Sinha 2000] the problem is solved and validated with measurements for several combinations of Re and Gr. In figure 3.9 these results are presented. The left-hand side shows the results obtained by [Sinha 2000] and the right side show the corresponding Comsol results. The verification results are satisfactory.

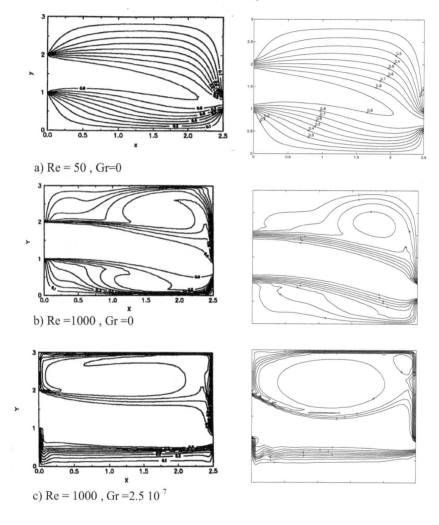

a) Re = 50 , Gr=0

b) Re =1000 , Gr =0

c) Re = 1000 , Gr =2.5 10^7

Figure 3.9 Dimensionless temperature contours comparison of the validated simulation results of [Sinha 2000] (left hand side) with the Comsol results (right hand side) for the 2D-airflow problem.

3.3.3 Discussion on reliability

The test cases in this Section show that Comsol is very reliable even for a highly non-linear problem such as convective airflow. Furthermore, for all the simulation results presented in this chapter, the default mesh generation and solver are used. So good results can be obtained without a deep understanding of the gridding and solving techniques. The simulation times (based on processor Pentium 3, 500 MHz) ranged from fast (order ~seconds) for linear problems such as the 2D thermal bridge example, medium (order ~minutes) for non linear problems in the coefficient form such as the 1D moisture transport example to high (order ~hours) for highly non linear problems in the general form such as the airflow problem.

3.4 3D combined heat and moisture transport

In this Section the challenging problem of simulating combined heat and moisture transport for a 3D geometry is presented. The problem is based on a 2D-problem [Kunzel 1996] but is now extended to a 3D problem. A brick specimen, initially dry and two of the flanks sealed, is placed with its lower surface about 1 cm under water. In figure 3.10, the geometry is presented.

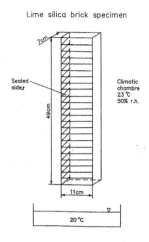

Figure 3.10 Geometry for the 3D combined heat and moisture transport.

The PDE model, boundary conditions and corresponding part of the code written in the coefficient form (3.1), are shown in figure 3.11.

$$\frac{\partial w}{\partial t} = \nabla \cdot (D_w(w) \nabla w + D_F(T, w) \nabla T)$$

$$\frac{\partial T}{\partial t} = \nabla \cdot (D_T(w) \nabla T)$$

$$w(t = 0) = w0, \quad T(t = 0) = T0,$$

sealed sides : $\nabla T = \nabla w = 0$

open sides : $\nabla T = A(T - Ti), \quad \nabla w = B(w - wi)$

bottom : $w(z = 0) = w\,max, \quad T(z = 0) = T0$

```
fem.dim=3;
fem.form='coefficient';
fem.equ.da={{1; 1}};
fem.equ.c={ 'Dw(w)'      'DF(T,w)'   0         'DT(w)'};
fem.bnd.g={ {0    0}     {B*wi A*Ti } {0 0} {0 0} {0 0} {0 0}};
fem.bnd.q={ {0    0}     {B    A  }  {0 0} {0 0} {0 0} {0 0}};
fem.bnd.h={ {1    1}     {0    0}    {0 0} {0 0} {0 0} {0 0}};
fem.bnd.r={ {wmax T0}    {0    0}    {0 0} {0 0} {0 0} {0 0}};
```

Figure 3.11 The PDE model and the corresponding Comsol code for the 3D combined heat and moisture transport.

In the PDE model 3 material properties functions are defined: Dw(w), DF(T,w) and DT(w). These functions are calculated using the material properties presented in [Kunzel 1996] and are shown in figure 3.12

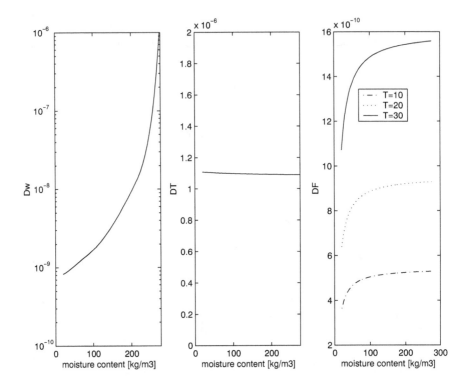

Figure 3.12   The material properties for the PDE coefficients Dw(w), DT(w)   and DF(T,w) based on [Kunzel 1996].

The results of the simulated moisture distribution initially and after 5, 10 and 70 days are visualized by slices and planes with equivalent moisture content, in figure 3.13[*].

---

[*] The reader should notice that the 2D validation data provided by [Kunzel 1996] can not be used in this case because we modeled (another) 3D case

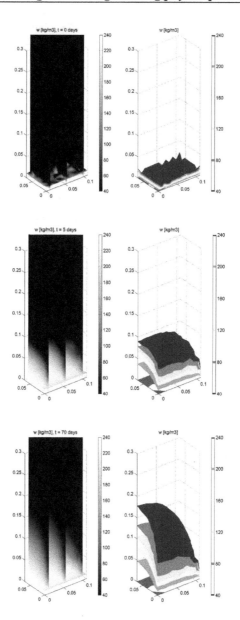

Figure 3.13   3D moisture distribution profiles for 0 days (top), 5 days (middle) and 70 days (bottom). Left hand side shows vertical slices and right hand side shows iso surfaces.

The extension of 3D modeling with different materials is left over for future research.

3.5 CONCLUSIONS

Comsol is evaluated as a solver for building physics problems based on partial differential equations (PDEs). Typical building physics problems such as moisture transport in a porous material, dynamic airflow and combined heat and moisture transport are relatively easy to model. The simulation results of the models used including 1D, 2D geometries show a good agreement with measurements[*]. The Comsol software is written in the MatLab environment and therefore it is possible to make use of the visualisation tools, toolboxes and all other programs written in MatLab. The evaluation illustrates the powerful and flexible nature of Comsol for solving scientific and engineering building physics problems.

REFERENCES

COMSOL 2006 http://www.comsol.com

Mathworks, 1998 Inc. MatLab Manual, Version 5.3, Reference Guide

Brocken, H., 1998, Moisture transport in brick masonry, Ph.D. thesis, Eindhoven University of Technology

Sinha, S.L., Arora, R.C. & Roy, S., 2000, Numerical simulation of two-dimensional room air flow with and without buoyancy, Energy and Buildings vol32., pp121-129

Kunzel, H.M., 1996 , IEA Annex 24 HAMTIE, Final Report - Task 1

---

[*] The 3D application is in fact a demonstration that Comsol can be used to model and simulate these types of problems.

Chapter 4

Integrated building physics simulation with Comsol/SimuLink/MatLab

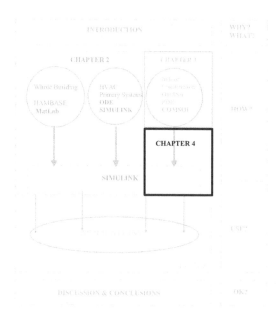

This Chapter covers the integration of PDE models into SimuLink. The combined MatLab toolboxes Comsol and SimuLink are evaluated as solvers for problems based on partial differential equations (PDEs). The following integrated building models are discussed: 1) A detailed case study on the combination of an indoor airflow model in Comsol connected to an on/off controller in SimuLink. In this case, the code of a Comsol model is implemented in the discrete section of a S-Function of SimuLink, so that the specialized solvers of Comsol can be used. 2) Other developed integrated building physics models in SimuLink are briefly discussed. These include a) a multi zone Heat Air and Moisture (HAM) indoor climate model, b) integration of this model with a thermal bridge model and controller and c) a 2D model of convective airflow around a convector. At the conclusion the applications and the possible benefits for scientists, engineers and designers are discussed.

(Section 4.1 – 4.3 from A.W.M. van schijndel published in proceedings of the 8 TH International IBPSA Conference, 2003, Vol. 3 pp. 1177- 1184;

Section 4.4 – 4.5 from A.W.M. van schijndel published in Building Serv. Eng. Res. Technol. 2003, 24 Vol 4 pp. 289 – 300)

4.1 INTRODUCTION

Many scientific problems in building physics can be described by PDEs. There are a lot of software programs available in which one specific PDE is solved. They are developed in order to get the simulation results in a short time and are most often emphasized on the simplicity of input of data, e.g. geometrical data. A disadvantage is that they often are not very flexible when the user wants to change or combine models. Also they most often act as black boxes.

Another category of software like Comsol [COMSOL 2000], is developed special for solving PDEs where the user in principle can simulate any system of coupled PDEs (for example the Navier Stokes equations and combined heat and moisture transport). The user can focus on the model (PDE coefficients on the domain and boundary) and does not have to spend much time on solving and visualization. Comsol is a MatLab Toolbox [Mathworks 1998] and has facilities to export models to SimuLink [Mathworks 1999] (Please note that the given references on Matlab, Comsol, and SimuLink are far from complete. They are intended as a starting point for further reading). The SimuLink simulation environment is flexible so that already developed building and HVAC models, like for example SIMBAD [CSTB 2001] can be coupled with the Comsol models. This combination can be very interesting for the following application areas:

* Science. Scientists can fully concentrate on the physics behind the models and the validation of models.

* Engineering. Engineers can combine or adapt already developed models for solving typical engineering problems using SimuLink. Furthermore, it is easy to create Graphical User Interfaces (GUI) of the input of SimuLink models, so engineers can develop user-friendly applications for end-users.

* Design. Designers only have to concentrate on the parameters of the models and the outcome of the simulation results when using user-friendly developed applications in SimuLink.

The further contents of the chapter is described now. Section 4.2 shows a complete example of an airflow problem, including the Comsol code and validation results. In

Section 4.3, a detailed case study on the integration of the presented Comsol airflow model in SimuLink is discussed. Section 4.4 summarizes other developed integrated building physics models.

4.2 A COMPLETE EXAMPLE

The test cases of Section 3.3.2 show that Comsol is very reliable even for a highly non-linear problem such as convective airflow. Furthermore, for all the simulation results presented in this chapter, the default mesh generation and solver are used. So good results can be obtained without a deep understanding of the gridding and solving techniques. The simulation time (based on processor Pentium 3, 500 MHz) are of order ~hours for highly non linear problems in the general form such as the airflow problem. Complete mfiles, results and movies can be found at [Comsol 2006].

Some limitations on Comsol CFD modeling

- 3D modeling is possible, however for practical usage, it consumes too much simulation time.
- k-ε is the only present turbulence model at this moment, other models are not developed yet.

Benefits of Comsol modeling

Scientists who want to create their own models can benefit from the combined PDEs modeling, the compact code, the easy adaptation of models, the state of the art solvers and the graphical output. Engineers can easily use the already developed models for other geometries and boundary conditions. The Comsol models can be exported and connected with MatLab/SimuLink models, creating a flexible simulation environment for combined PDE and ODE (ordinary differential equation) based models (see the following Section).

4.3 AIRFLOW AND CONTROLLER

This Section shows in detail how an airflow model of Comsol can be coupled to a controller of SimuLink. Comsol has standard facilities to export models to SimuLink [COMSOL 2006]. However, a standard export of the airflow model to SimuLink is not very practical because such a non-linear model can not be simulated efficiently by SimuLink (the standard solvers of SimuLink can not handle such a problem very well). A possible solution to this problem is to implement the Comsol code in the discrete section of a SimuLink S-function. The S-function solves each time step (in this case 1 sec) an airflow problem using the Comsol solver. After each time step the solution is exported. Dependent on the controller, different boundary values can be applied. The use of the implementation of Comsol in a S-function is demonstrated in the following example. Again the Comsol model of Section 3.3.2 is used. The airflow at the inlet is now controlled by an on/off controller with hysteresis (Relay) of SimuLink. If the temperature of the sensor is above 20.5 °C the air temperature at the inlet is 17 °C if the air temperature is below 19.5 °C the inlet temperature is 22 °C. Initially the inlet temperature is 18 °C. Figure 4.1 through 4.4 give an overview of the results.

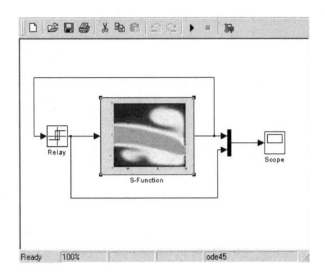

Figure 4.1 The Simulink model including the S-function of the Comsol model and controller (Relay)

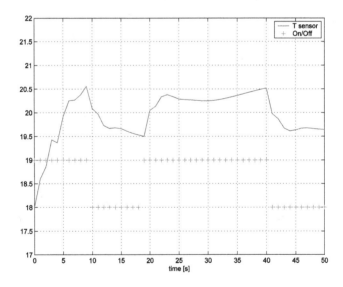

Figure 4.2 The temperature of the sensor (-) and the output of the on/off controller (+) versus time

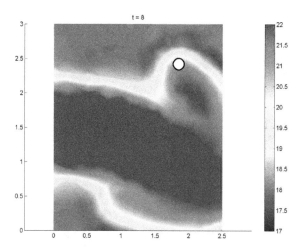

Figure 4.3 The air temperature distribution at 8 sec. (hot air is blown in) (o = sensor position).

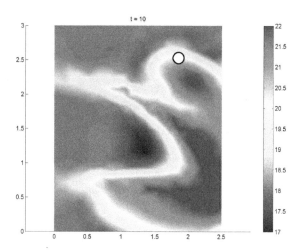

Figure 4.4 The air temperature distribution at 10 sec. (cold air is blown in) (o = sensor position).

Figure 4.5 shows details of how a S-function should be programmed for the use with this Comsol model and solver.

```
function [sys,x0,str,ts] = dsfun(t,x,u,flag)
% t: time; x = state vector ; u = input vector; % flag = control parameter
global OutT Outsol
% Comsol model geomtry, mesh, PDE coefficents and      boundary conditions
switch flag,
    case 0,
    [sys,x0,str,ts] = mdlInitializeSizes(fem,tstap,u0);% Initialization %
    case 2,
    sys = mdlUpdate(t,x,u,fem,tstap); %update discrete part
        case 3,
    sys = mdlOutputs(t,x,u,fem,tstap); %produce output
    case 9,
    sys = []; %terminate
  end %end dsfunc
function [sys,x0,str,ts] = mdlInitializeSizes(fem,tstap,u0)
global OutT Outsol
% calculated the next values needed for SimuLink
% calculated initial values for time OutT(1) and
% initial values for solution Outsol(1)
% x0 are all values of the temperature on the mesh
OutT(1)=0;   Outsol(1)=x0;
function sys = mdlUpdate(t,x,u,fem,tstap)
global OutT Outsol
if u(1)>0.5 % if dimensionless Temperature of sensor > 0.5 then
    fem.bnd.r={    {'-u1'; '-u2' ;0 ; '0-u4'} ... %boundary value cold air
else
    fem.bnd.r={    {'-u1'; '-u2' ;0 ; '1-u4'} ... % bounadry value hot air
end
tijd=[t; t+tstap/2; t+tstap]; %calculate simulatetime for Comsol model
fem.sol=femtime(fem,'atol',{{.1,.1,Inf,.1}},'ode','fldae',...
        'sd','on','report','on','tlist',tijd,'init',x); %simulate airflow
OutT=[OutT  (t+tstap)]; % update time
Outsol=[Outsol   fem.sol.u(:,ntijd)]; % update solution
function sys = mdlOutputs(t,x,u,fem,tstap)
global OutT Outsol
Points=[69 84 101 167 188 211 259 261]; % location of sensor in mesh
sys=mean(x(Points)); % output
```

Figure 4.5 Example of the implementation of a Comsol model in the discrete section of a S-function of SimuLink

More Details about S-functions can be found in [Mathworks 1999]. The above results show that also highly non-linear models solved with Comsol can be exported to SimuLink by writing an appropriate S-function. The complete mfiles, results and movies can be found at [Comsol 2006].

Benefits:

Engineers can use this model to study the effect of controller type and settings on the indoor airflow and temperature distribution.

4.4 OTHER DEVELOPMENTS

In this Section some other recent developments in the area of integrated building physics with Comsol/ SimuLink/MatLab are discussed. In order to integrate HAM models, a multi zone indoor climate model in SimuLink has been developed:

4.4.1 2D Convective airflow around a convector

The objective of this project was to investigate the use of Comsol in case of a pure convective airflow. The practical application was to find the minimal surface temperature of a convector to compensate cold airflow due to a cold surface. The same model as in Section 3.2 has been used, only the geometry and boundary conditions are adapted. In figure 4.6 the geometry is shown:

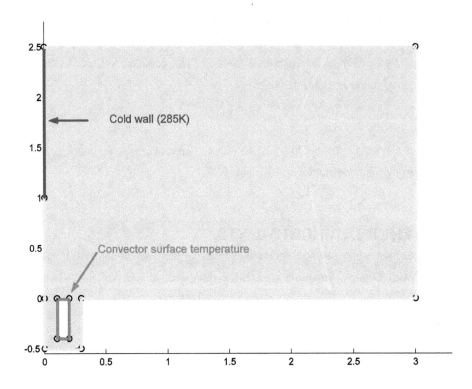

Figure 4.6 The geometry of the convector, cold surface and room (3m x 2.5 m).

In figure 4.7 and 4.8 the simulation results are shown for an equivalent convector surface temperature of 25 °C and 37.5 °C respectively. The initial room temperature is 20 °C.

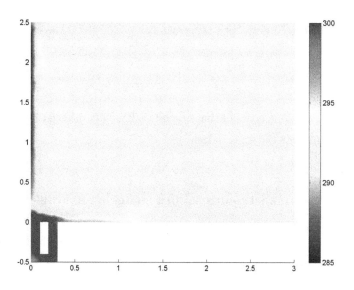

Figure 4.7 The room temperature (in K, surface convector temperature is 25 °C)

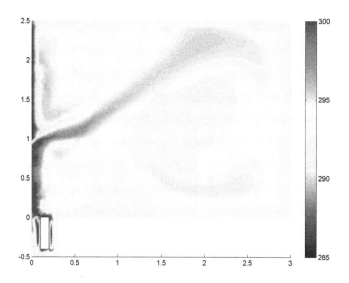

Figure 4.8 The room temperature (in K, surface convector temperature is 37.5 °C)

Figure 4.7 and 4.8 show that a mean convector surface temperature of 25 °C does not compensate the cold airflow, but a temperature of 37.5 °C does. The model is validated by measurements for the given geometry [van Schijndel & van Aarle 2000].

Benefits
Designers can use this model to simulate the airflow with different temperatures and sizes for the cold wall and the convector.

4.4.2 A Comsol model connected to a model in SimuLink

The objective of this project is how to connect a Comsol model to a model in SimuLink. The thermal bridge model of Section 3.2 has been exported using the standard export facilities of Comsol and placed into the room model. In figure 4.9 the complete model is shown:

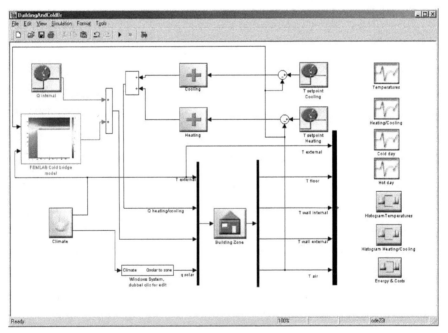

Figure 4.9. The thermal bridge (Comsol) model connected with the room model.

The practical application[*] was to study the effect of lowering the air temperature during the night on the relative humidity near the thermal bridge. Figure 4.10 and 4.11 show the results:

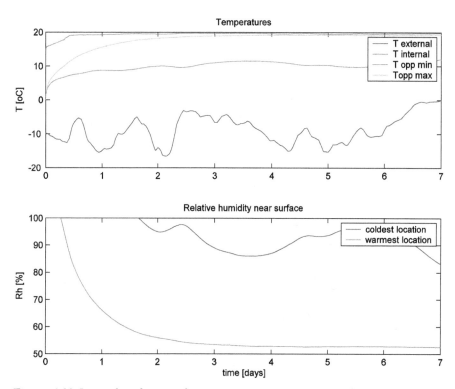

Figure 4.10 Internal and external air temperature, minimum and maximum surface temperature at the thermal bridge (top) and min./max. relative humidity near the thermal bridge (bottom) versus the time [days] with room temperature set point is held at 20 °C

---

[*] In this case, the indoor vapor pressure was fixed at 1170 Pa.

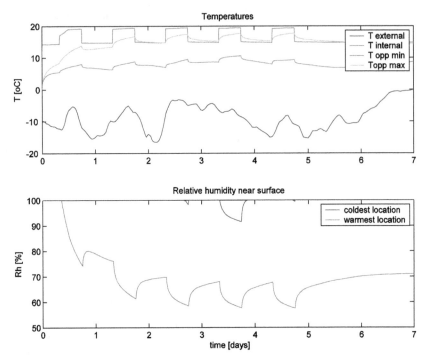

Figure 4.11 Internal and external air temperature, minimum and maximum surface temperature at the thermal bridge (top) and min./max. relative humidity near the thermal bridge (bottom) versus the time [days] with room temperature set point lowered to 15 °C during the night.

The effect of lowering the air temperature during the night on the relative humidity near the thermal bridge is clear. The model is used as a demonstration model for students to simulate the interaction between building construction (the thermal bridge), the indoor climate and controller settings (room model).

This project showed that a linear model in Comsol could easily be exported to SimuLink. The simulation time stays within limits (above simulation takes about 3 minutes on a Pentium 600MHz). Note that not only detailed temperatures of the thermal bridge are simulated but also the heat flow to the thermal bridge from the room.

Benefits

Engineers who have some experience with SimuLink can combine Comsol models with their own models and controllers in SimuLink.

4.5 CONCLUSIONS

The combination Comsol/SimuLink/Matlab is evaluated as solver for HVAC problems based on partial differential equations (PDEs). An example of the implementation of a typical HVAC modeling problem such as a dynamic airflow in combination with a controller is demonstrated. The simulation results of the presented airflow model agree reasonably well with measurements. Other presented models show that the combination Comsol/SimuLink/MatLab is a powerful and flexible environment for modeling and solving HVAC problems.

Benefits for scientists are:

> * The wide application area. The software is designed to simulate systems of coupled PDEs, 1D, 2D or 3D, non-linear and time dependent.
>
> * There is a clear relation between the mathematical model and (compact) program code in terms of specified PDE coefficients and boundary conditions.
>
> * The easy-to-use default gridders and solvers give accurate solutions.
>
> * The graphical output capabilities.

Benefits for engineers are:

> * Already developed SimuLink models and controllers can be connected with Comsol models.
>
> * It is relatively easy to combine or adapt models for solving typical engineering problems.
>
> * It is relatively easy to create Graphical User Interfaces (GUI) of the Simulink models.

Benefits for designers are:

The Comsol/SimuLink/MatLab environment facilitates state of the art PDE modeling with user-friendly Graphical User Interfaces.

REFERENCES

COMSOL, 2000, COMSOL Version 2.0 pre, Reference Manual, September 2000

COMSOL, 2006 http://www.comsol.com

CSTB, SIMBAD, 2001, Building and HVAC Toolbox, Reference Manual, November 2001

Schijndel, A.W.M. en M.A.P. van Aarle, 2000, (Dutch) Onderzoek naar de werking van convectoren bij lage temperaturen, TU Eindhoven, Intern FAGO rapport

Sinha, S.L., Arora, R.C. & Roy, S., 2000, Numerical simulation of two-dimensional room air flow with and without buoyancy, Energy and Buildings vol32., pp121-129

The Mathworks, 1998 Inc. MatLab Reference Manual, Version 5.3.

The Mathworks, 1999, Inc. Simulink Reference Manual, Version 3.

PART II. DESIGN

THE SIMULATION ENVIRONMENT
AS A TOOL FOR DESIGN

Preface

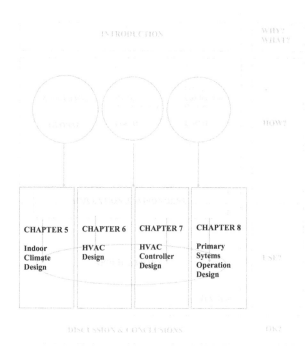

Part II presents the case studies, focussing on the *use* of the simulation environment for performance-based design. The order of applications is based on an increase of complexity of the building systems and operation, i.e. the first application contains relative simple systems and the last application has the most complex systems and operations.

Chapter 5 presents a case study on the performance-based design for the indoor climate of a monumental hall. The aim of this study is to classify and evaluate the indoor climate performances for the preservation of the monumental interior.

Chapter 6 provides a case study on the performance-based design of a HVAC system and controller of a museum. The study evaluates design solutions for the HVAC system and showcases for the preservation of paper fragments

Chapter 7 presents a case study on the optimal operation of the climate control of a monumental church. Several operation strategies based on temperature and relative humidity change rates control are evaluated for best preservation of the monumental wooden organ.

Chapter 8 presents a case study on the optimal operation of a complex hospital power plant. A model is used to calculate the energy loss and costs for different operation strategies.

Chapter 5

Indoor climate design for a monumental building with periodic high indoor moisture loads

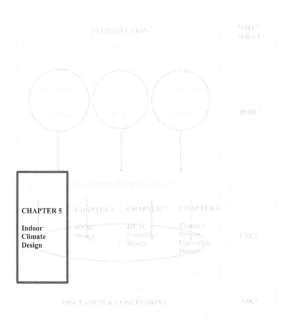

This Chapter presents a case study on the performance based design for the indoor climate of a monumental building with periodic high indoor moisture loads. Several scenarios of the past performance and new control classes are simulated and evaluated. The results include the influence of hygric inertia on the indoor climate and (de)humidification quantities of the HVAC system. It is concluded that: (1) The past indoor climate can be classified as ASHRAE control C with expected significant occurrences of dry (RH below 25%) and humid (RH above 80%) conditions; (2) ASHRAE control C is not suitable for the new hall. The climate control classification for the new hall ranges from B to AA.; (3) The demands on the HVAC system to facilitate pop concerts in the new hall are 40 kW heating power, between 100 and 200 kW cooling power, between 40 and 80 kW humidification power and 125 kW dehumidification power; (4) In case of control class AA, placing additional hygroscopic material has no significant effect. In case of control class B, the placing of additional moisture buffering material (5 air-volume-equivalents) does not decrease the (de)humidification power but decreases the (de)humidification energy by 5%.

(A.W.M. van Schijndel published in Proceedings of the 26TH AIVC Conference in Brussel 2005 , pp301-313)

5.1 INTRODUCTION

A monumental theatre, formerly used as a cinema, is renovated. The interior of the theatre, containing monumental paintings, wood and plaster, is well preserved. It is concluded that heating of the building in the past had no significant impact on the monumental interior. A new destination of the renovated theatre is to facilitate (pop) concerts. This will cause a much higher indoor moisture load as before. An important demand is that the indoor climate may not deteriorate the interior compared to the past situation. Currently the building is already under renovation, so measurement of the original situation was not possible. Therefore simulation is the only tool that can be used to check the indoor climate performances of the past situation and of new designs. The objectives of this paper are twofold: (I) Demonstration of the use of a new simulation and visualization tool by characterization of the past performance of the indoor climate; classification of the new climate control and calculating the required HVAC capacities. (II) A preliminary study of the effect of moisture buffering on the energy needed for (de)humidification by the HVAC system. The outline of the paper is as follows: Section 5.2 presents background information on the building and its monumental interior, performance criteria for the indoor climate and the used simulation tool HAMBase. Section 5.3 provides simulation results of the past indoor climate of the Luxor hall and new designs with and without placing additional hygroscopic material, using the new visualization chart. In Section 5.4, a discussion on the simulation results is presented. Section 5.5 shows the conclusions.

5.2 BACKGROUND

The current building and its monumental interior

The Luxor theatre, designed by Willem Diehl, was built in 1915 in Arnhem (Netherlands). The main hall (volume about 1800 m³) has no windows. The outdoor walls are made of brick (0.8m) with air gap. The roof is made of tiles. It's monumental interior consists of a wooden stage frame, gypsum/wooden ceiling artefacts and paintings on wood and paper. Figure 5.1 provides an impression of the building and the hall.

Figure 5.1 Left: The Luxor Theatre in 1931; Right top: The monumental wooden stage frame; Right bottom: Gypsum/wooden ceiling artefacts.

The performance criteria and classification of the indoor climate
The performance criteria for the indoor climate are based on control classes [ASHRAE 2003]. Details can be found in Section 5.3.

The modeling tool HAMBase
HAMBase is used as simulation tool (see PART I).

5.3 SIMULATION RESULTS

Climate Evaluation Chart (CEC)

All simulated scenarios are presented by Climate Evaluation Charts (CEC). Because there is a lot of information in the chart, an explanation of the CEC is given first. Figure 5.2 presents an typical example of a CEC. The interpretation of the chart is explained below (the data itself are not important at this moment)

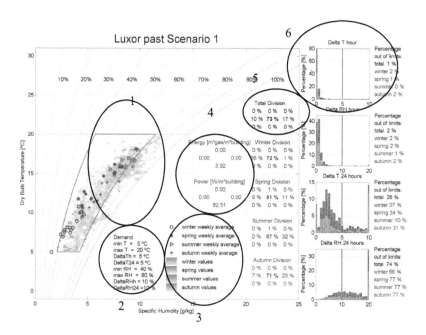

Figure 5.2 Example CEC.

The background of the chart is a standard psychometric chart for air, with on the horizontal axis: the specific humidity, on the vertical axis: the temperature and curves for the relative humidity. Area 2 shows the demanded performance (demands) on: (1) indoor climate boundaries: minimum and maximum temperature and relative humidity (min T, max T, min RH and max RH) and (2) indoor climate change rate boundaries: maximum allowed hourly and daily changes in temperatures and relative humidities (DeltaTh, DeltaT24, DeltaRHh, DeltaRH24). Area 1 shows the indoor climate boundaries and the simulated indoor climate of a building exposed to a (Dutch standard test reference) year. The simulated climate is presented by seasonal (Spring from March 21 till June 21, etc.) colours representing the percentage of time of occurrence and seasonal weekly averages. The colours visualize the indoor climate distribution. For example, a very stable indoor climate produces a narrow spot, in contradiction to a free floating climate which produces a large 'cloud'. Area 3 provides the corresponding legend. Area 5 shows the total percentage of time of occurrence of areas in the psychometric chart (9 areas). In this example 73% of the time the indoor climate is within the climate boundaries; The area to the left (too dry) occurs 10% of the time, the area to the right (too humid) occurs 17% of the time. The climates in the other 6 regions do not occur. Below area 5 the same information can be found for each season. Area 4 shows the energy consumption (unit: m^3 gas / m^3 building volume) and required power (unit: W / m^3 building volume) used for heating (lower), cooling (upper), humidification (left) and dehumidification (right), assuming 100% efficiencies. In this example the energy amount is 3.92 m^3 (gas / m^3 building volume) and required power is 82.51 (W / m^3 building volume) used for heating. Cooling, humidification and dehumidification are zero in this example. Area 6 presents the occurrence (in percentage of time) outside the climate change rate boundaries. In the example the demand of maximum allowed hourly change of temperature of 5 ($^\circ$C/hour) is shown as a blue line. The distribution per season is provided together with the percentage of time of out of limits. In this example, area 6 shows that only 1% of the time, the hourly temperature change rate is out of limits. This is also specified for each season. Below area 6 the same can be found for the other climate change rate boundaries. All simulation results will be presented below.

Simulated scenarios

All simulations are performed using a test reference year for the Netherlands. The past performance is simulated using two scenarios, representing periodic and intensive use of the hall in the past. The goal of these scenarios is to simulate worst-case conditions.

The results are used for comparison with new control class designs. The future use of the hall includes (pop) concerts. This will cause a much higher moisture load in the hall. This effect is included in all future scenarios. The simulated control class scenarios [ASHRAE 2003] start with the lowest class, representing the same type of HVAC system as in the past and end with the highest class, representing precision control. Furthermore, extra scenarios are included with the use of additional hygroscopic material. The goal of these scenarios is to study the effect of moisture buffering on the indoor climate, energy consumption and required power. After presenting all scenarios the results will be discussed in Section 5.4.

The hall used as cinema

The hall was heated and ventilated by an air system. During a performance (duration 4 hours, 200 sitting people producing 17 kW heat and 2.5 g/sec moisture) the system provided an estimated ventilation rate of 10 ACH. During the day (duration 10 hours, 2 people) this was estimated as 2 ACH and 1 ACH for the rest (including infiltration). In order to estimate the past indoor climate, two extreme scenarios are simulated. The assumption is made that the past indoor climate was somewhere between the next two past scenarios: (1) periodic use of the hall: 2 times a week a performance, the hall is only heated during these performances. Furthermore a minimum air temperature of 5 °C is maintained. (2) intensive use of the hall: 6 times a week a performance, the hall is also heated during the day. A minimum air temperature of 10 °C is maintained. The results are presented in figure 5.3 and 5.4.

All *new designs* have to facilitate: (a) 2 concerts a week (duration 4 hours, 700 moving people producing 63 kW heat, 27 g/sec moisture) with a designed ventilation rate of 10 ACH; (b) 3 meetings a week (duration 4 hours, 100 people producing 8 kW heat, 1 g/sec moisture) with a designed ventilation rate of 10 ACH; (c) During the day (duration 10 hours, 10 people, 2 ACH) and 1 ACH for the rest (including infiltration).

79

A control class C design

First, a control class C design will be evaluated. Class C, defined as 'to prevent all high risk extremes', usually consists of basic heating cooling and ventilation. Two scenarios are presented: (1) HVAC system: heating, cooling and ventilating, without additional hygroscopic material, (2) HVAC system: heating, cooling and ventilating, with additional hygroscopic material. The results are presented in figure 5.5 and 5.6 (Note: There are no 'percentages out of limits' presented in these CECs, because (ASHRAE) control class C does not specify a limitation of the allowable change rates).

A control class B design

Also, a control class B design will be evaluated. Class B, defined as 'precision control, some gradients plus winter temperature setback', is usually a HVAC system, including cooling and (de)humidification . Two scenarios are provided: (1) HVAC system: climate control, without additional hygroscopic material. (2) HVAC system: climate control, with additional hygroscopic material. The demanded performance and results are presented in figure 5.7 and 5.8.

A control class AA design

Finally a control class AA design will be evaluated. Class AA, defined as 'precision control, no seasonal changes', is usually a high-tech HVAC system, including cooling and (de)humidification. One scenario is presented: HVAC system: climate control. The demanded performance and results are presented in figure 5.9.

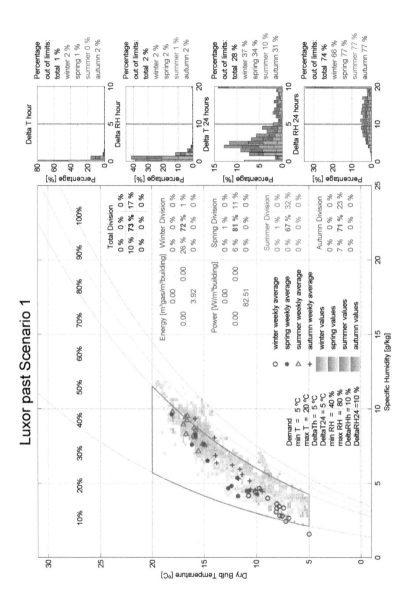

Figure 5.3 The CEC of Scenario 'periodic use of the hall'

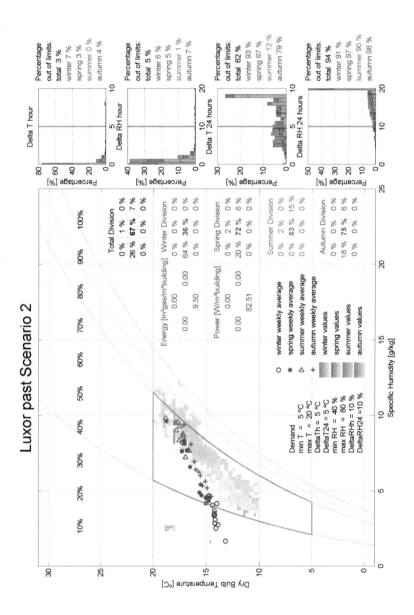

Figure 5.4 The CEC of Scenario 'intensive use of the hall'

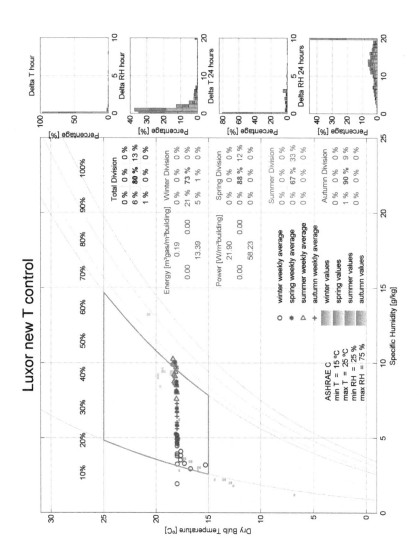

Figure 5.5 The CEC of Scenario 'control class C design without additional hygroscopic material

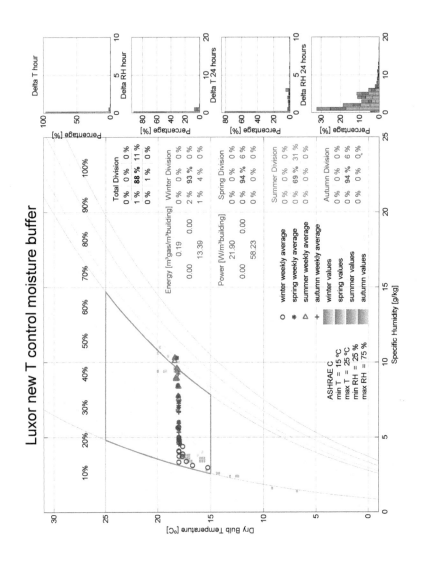

Figure 5.6 The CEC of Scenario 'control class C design with additional hygroscopic material

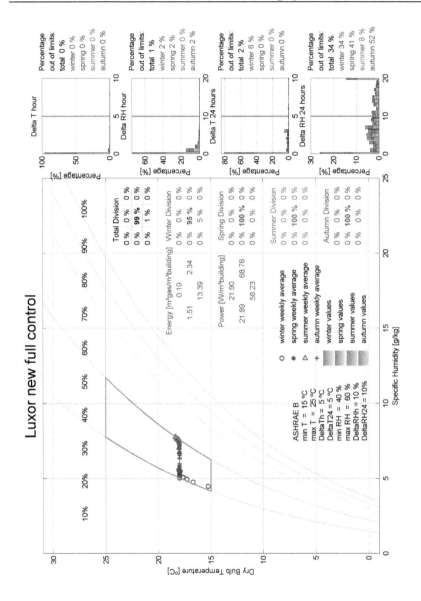

Figure 5.7 The CEC of Scenario 'control class B design without additional hygroscopic material

85

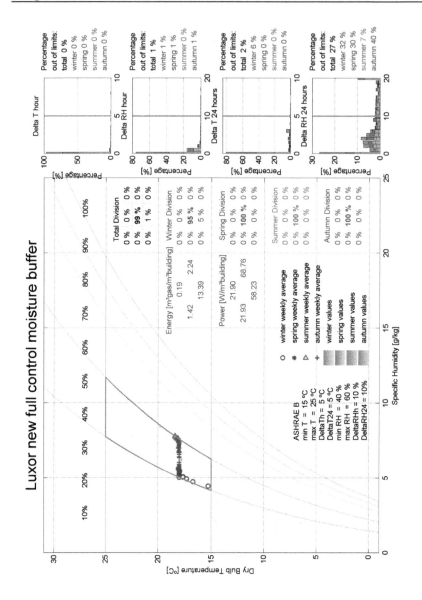

Figure 5.8 The CEC of Scenario 'control class B design with additional hygroscopic material

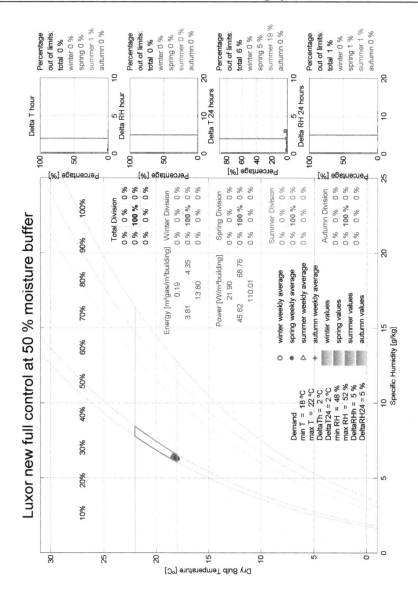

Figure 5.9 The CEC of Scenario 'control class AA design.

5.4 DISCUSSION OF THE RESULTS

5.4.1 Evaluation of the scenarios

Past performance

Both past performance scenario results show high RH change rates (up to 94% of time out of limits) and high humidities (up to 17% of time is too humid) . Although this did not lead to visible damage[*] to the monumental interior so far, it should be reduced to prevent future damage. Furthermore, the other scenario in figure 5.4, results in low humidities (up to 26% of time is too dry). Both problems should be prevented in the future.

Control class C design

This design contains basic heating cooling and ventilation. Figure 5.5 shows that this is not an acceptable design because: (a) during the winter the RH is 21% of the time below 25% and (b) 40% of the time the daily RH fluctuation is above 20%. Figure 5.6 shows that both problems are solved if a very large amount of additional moisture buffering material (100 air-volume-equivalents) is placed. The amount of additional moisture buffering material is expressed in air-volume-equivalents (of the 1800 m^3 hall). This means 1 air-volume-equivalent (at 20 °C/50%) equals 14 kg of moisture. If wood fibreboard is selected as buffering material, 1 m^3 wood can buffer 12 kg moisture (daily change rate of Rh 20%, density wood = 300 kg/m^3, (de)sorption = 0.04 $kg_{moisture}/kg_{wood}$). 100 wood slices of 1 cm x 1m x 1m separated by 1 cm air gap fills 2 m^3 of space. Thus 100 air-volume-equivalent takes 230 m^3 of space. However, due to the monumental roof and walls, it will be very difficult to place such amount of buffering material. This means that moisture control will be inevitable.

Control class B design

This design contains full climate control with RH between 40% and 60%. Figure 5.7 shows that the only problem left is that still 34% of the time the daily RH fluctuation is above 20%. Figure 5.8 shows that if a reasonable amount of additional moisture buffering material (5 air-volume-equivalents, taking 12 m^3 of space) is placed, then: (a)

---

[*] No visible damage does not mean that there is no damage at all.

it has only small effects on the humidification energy (decrease of 6%) and dehumidification energy (decrease of 4%); (b) it has no significant effect on the (de)humidification power and (c) the daily RH fluctuations remain too high. Note that these climate conditions are expected to be better than the past performance. One could argue that the monumental interior has not deteriorated in the past, so it probably will not deteriorated in better climate conditions. In this case control class B would be appropriate. But one can also argue that it is pure luck that the monumental interior has not deteriorated so far. In this case control class B would not be appropriate and a more stringent control class is required.

Control class AA design

This design contains full climate control with RH between 48% and 52%. This means there is almost no moisture buffering due to the steady RH. Figure 5.9 shows that the climate meets the demanded performance very well[*]. The disadvantages compared to the previous class B design are: (a) much higher humidification (increase of 270%) and dehumidification energy consumptions (increase of 200%) and (b) a significant effect on the humidification power (increase of 200%) and cooling power (increase of 190%).

5.4.2 Evaluation of the moisture buffering effects on the HVAC performance

From figure 5.7 and 5.8 it follows that the humidification energy drops from 1.51 to 1.42 [m^3 gas / m^3 building volume] and the dehumidification energy drops from 2.34 to 2.24 [m^3 gas / m^3 building volume] by placing 5 air-volume-equivalents of moisture buffering material. The next figure presents the (de)humidification energy as a function of the additional buffering material in more detail.

---

[*] This study focusses on the indoor climate. Further research is recommended to study the impact of this control class on the (monumental) constructions.

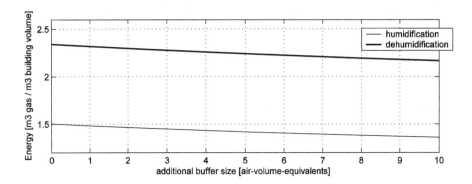

Figure 5.10 The (de)humidification energy as a function of the placed air-volume-equivalents.

Figure 5.10 shows an almost linear decrease of 0.01 [m³ gas / m³ building volume] per air-equivalent additional buffering material.

5.5 CONCLUSIONS

Revisiting the objectives of Section 5.1, the conclusions are:

In relation to the use of the new simulation and visualization tool: (1) The past indoor climate can be classified as ASHRAE control C with expected significant occurrences of dry (RH below 25%) and humid (RH above 80%) conditions. (2) ASHRAE control C is not suitable for the new hall. The climate control classification for the new hall ranges from B to AA. Further research in the form of detailed HAM transport calculations in the (monumental) constructions are needed to select the appropriate class. (3) The demands on the HVAC system to facilitate pop concerts in the new hall are 40 kW heating power, between 100 and 200 kW cooling power, between 40 and 80 kW humidification power and 125 kW dehumidification power. All these quantities are based on 100% process efficiencies.

Considering the effect of moisture buffering on the (de)humidification energy of the HVAC system, the conclusions are: (4) In case of control class AA, placing additional hygroscopic material has no significant effect. In case of control class B, the placing of additional moisture buffering material (5 air-volume-equivalents) does not decrease the (de)humidification power and it decreases the (de)humidification energy by 5%.

ACKNOWLEDGEMENT

Rogier Lony provided input data for the models. Marco Martens contributed to the development of the CEC. Both are greatly acknowledged by the author.

REFERENCE

ASHRAE, 2003. Heating, ventilating and Air-Condioning Applications, Chapter 20, museums, libraries and archives. ISBN 1-883413-72-9,

Chapter 6

Application of an integrated indoor climate & HVAC model for the indoor climate performance of a museum

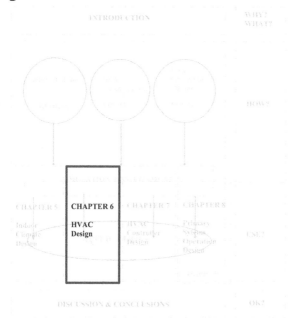

This chapter presents a case study on the performance based design of a HVAC system and controller of a museum. A famous museum in the Netherlands has reported possible damage to important preserved wallpaper fragments. The paper provides an evaluation of the current indoor climate by measurements, showing that the indoor climate performance does not satisfy the requirements for the preservation of old paper. To solve this problem, possible solutions are evaluated by simulations using integrated heat air & moisture (HAM) models of respectively: the indoor climate, the HVAC system & controller and a showcase. The presented models are validated by a comparison of simulation and measurement results. An integrated model consisting of all different models is applied for the evaluation of a new HVAC controller design and the use of a showcase. The results are discussed.

(A.W.M. van Schijndel & H.L. Schellen published in proceedings of the 7ᵀᴴ Symposium on building Physics in the Nordic countries 2005 , vol 2; pp1118-1125)

6.1 INTRODUCTION

In general, the aim of museums is to exhibit artefacts in its original state as long as possible. The climate performance surrounding the preserved artefact is of great importance. Furthermore, if present, the heating, ventilation and air-conditioning (HVAC) system plays a dominant role on the indoor climate. A famous museum in the Netherlands has reported possible damage to important preserved wallpaper fragments. Preliminary measurements indicate that the indoor climate performance does not meet the criteria for preservation of wall paper. A solution is sought-after, given that the current HVAC system cannot be replaced (only small modifications are possible) and that the use of showcases, although not prohibited, is not preferred by the decision makers. This leads to the next questions: First, what are criteria of the indoor climate for preservation of wall paper? Second, is it possible to improve the indoor climate performance, by a new control strategy of the current HVAC system, in such a way, that a showcase can be avoided? Third, if not, can the problem be solved by using a showcase? Due to the preservation of the object, measurement is not an option to answer these key questions because it is not allowed to experiment with the HVAC system. Therefore simulation is the only option and an integrated indoor climate, HVAC and showcase model is needed. There is no such a model available. This leads to two more key questions: First, can we develop an integrated heat air & moisture (HAM)/HVAC system model capable of predicting the current indoor climate and the climate in a showcase? Second, can we improve the climate surrounding the object, using this model?

The aim of this chapter is to answer the key questions. The following methodology was used: (1) Reviews on the indoor climate criteria for preservation of wallpaper and on integrated indoor climate, HVAC and showcase models have been carried out. (2) The current indoor climate and HVAC performances were extensively measured. Data, measured by others, have been obtained for validating the showcase model. (3) Indoor climate, HVAC and showcase models were developed and validated. (4) An integrated model has been developed for the simulation of climate conditions near the object in case of a new HVAC controller design, with and without the use of a showcase.

The outline of the chapter is as follows: Section 6.2 presents review results on the optimal climate for the preservation of the object and the measured results of the actual indoor climate performance. Section 6.3 through 6.6 provide all modeling and simulation results by respectively HAM models of the indoor climate, HVAC system and showcase (Section 6.3), validation of the developed HAM models (Section 6.4) and applications of integrated HAM models (Section 6.5). Finally in Section 6.6, the key questions will be revisited and discussed.

6.2 THE CURRENT INDOOR CLIMATE PERFORMANCE

6.2.1 Review on climate recommendations for (wall) paper

[Wijffelaars & Zundert 2004] reviewed several recommendations for the preservation of paper from literature. A short summary of this review is presented now. [Appelbaum 1991] recommends a relative humidity (RH) between 40 and 50%. [LCM Foundation, 2002] presents the next criteria for paper: (1) A RH between 48 and 55%. (2) A RH variation less than 3% per day. (3) An air temperature (T_a) between 16 and 18 $^{\circ}$C. (4) A T_a variation less than 2 $^{\circ}$C per hour and RH variation less than 3% per day. [Jutte 1994] recommends a RH between 48 and 55%. [Henne 1995] and [Johnson & Horgan 1979] both present criteria on storage conditions for paper. Their recommendations are almost the same: a RH between 45 and 60% and a mean T_a of 20 $^{\circ}$C. Furthermore [Wijffelaars & Zundert 2004] discussed their review results with an expert on paper preservation. It was concluded that the [LCM Foundation 2002] provided the best recommendations for the preservation of the wallpaper. The next Section presents the actual indoor climate conditions surrounding the wallpaper.

6.2.2 Measurements

We start with a short description of the situation. All important paper fragments are fixed at several walls in a single room. The window is orientated north-east. The room is permanently filled with about 10-20 persons during museum opening times. Information on the HVAC system is provided in Section 6.3. Figure 6.1 provides some information on the geometry and constructions:

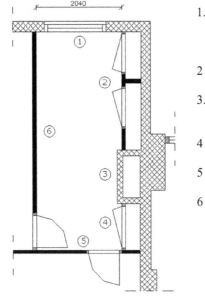

1. external wall, (2x2.7m²),
   plaster - 0.2m brick
   window, (1.1x1.7m²), Solar
   Gain Factor=0.25
2. external wall, (2.4x2.7m²),
   wood - air gap - 0.2m brick
3. internal wall, (1.6x2.7m²),
   0.1m brick- air gap - 0.2m
   brick
4. internal wall, (1x2.7m²),
   wood - air gap - 0.2m brick
5. internall wall, (2x2.7m²),
   wooden panels
6. internal wall (5x2.7m²),
   wooden panels

   ceiling, (5x2m²), wood

   floor, (5x2m²), linoleum -
   wood

Figure 6.1 The geometry of the room and material properties of the constructions

The measurements were carried out from June 2003 through February 2004. They include: (1) The T_a, RH, solar irradiance and rainfall of the external climate. (2) The T_a and RH at several places in the room and surrounding rooms. (3) The T_a, RH and mass flow of the air inflow to the rooms. Figure 6.2 presents the indoor climate conditions of the room that contains the wallpapers.

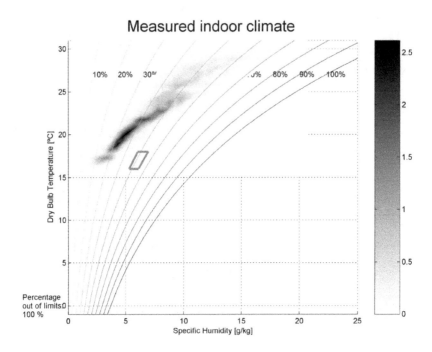

Figure 6.2 The Mollier frequency plot of the measured indoor climate for June 2003 through February 2004. The grey area shows the percentage of time for each state to occur (scale see the colour bar). Summation of the whole coloured area equals 100%. In blue the recommended area is provided.

We used a Mollier frequency plot for the visualisation of the measured indoor climate. It is very clear that the indoor climate does not satisfy the recommendations: First, from the Mollier frequency plot, it is concluded that 100% of the measured values are outside the recommended area (!). Second, from the variations plots, it is concluded that only the T_a change per hour is within the recommendations. Third, the RH change per day is almost 100% out of limits. As mentioned before, to solve this problem, simulation is the only option to study possible solutions of this problem. The next Section presents the used heat, air & moisture (HAM) models.

6.3 HAM MODELING AND VALIDATION

6.3.1   A short review on HAM modeling

(This short review, presented in the original paper, is omitted because it has already extensively been discussed in Section 1.2.)

6.3.2 The indoor climate and HVAC modeling

All models are developed using HAMBase [see PART I]. Figure 6.3 shows the input/output structure of the models and validation results.

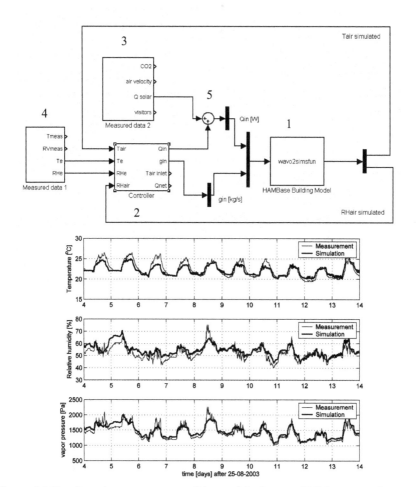

Figure 6.3 Top: Input/output structure of the models. Bottom: Validation results

The room model (1) has as inputs: heat flow Q_{in} and moisture flow g_{in} and output: T_a and RH. The HVAC system & controller model (2) has as inputs: indoor and outdoor climate and as outputs: heat flow and moisture flow. The measured outdoor climate is provided by (4). The heat flow to the room due to measured solar irradiation is provided by (3) and added to the heat flow of the HVAC model at (5). Bottom: measured and simulated room T_a, RH and vapour pressure.

The room, mentioned in Section 6.2.2, is modelled using HAMBase (see PART I). Furthermore, the HVAC system controller is modelled using the current control strategy for the incoming air temperature T_a and a constant airflow of 1000 m^3/hour. T_a is a function of the outdoor temperature (T_e) and indoor RH: T_a = $f(T_e,RH)$= max(17 ,20-$(T_e$+10)/3)+min (10, (RH-40)/2.5). After completing the models with the necessary input data, the complete model was validated by measurements with quite satisfactory results (see figure 6.3).

6.3.3 The showcase modeling

[Wijffelaars & Zundert 2004] developed a HAM model for a showcase. The model will be discussed very shortly, to get an idea of the background. Figure 6.4 provides a schematic view of the modelled quantities, the moisture related model equations[*] and validation results, where m is the water vapour mass; T is temperature; $m_{max}(T)$ is maximum water vapour mass at temperature T; RH is relative humidity; β_{RH} surface coefficient of vapour transfer; δ vapour permeability coefficient, μd the vapour diffusion thickness of paper; $p_s(T)$ is vapour saturation pressure at temperature T; $dp_s(T)$ is air saturation pressure derivative at temperature T. The surface was 1m^2. Subscripts: *1-8* are locations presented at the bottom of figure 6.4. The showcase model was validated with data, measured by others. Unfortunately these data did not contain measured RH values inside the showcase. Therefore only the thermal part of the model

---

[*] The heat related model equations are based on a straight forward network of thermal resistances

is validated with satisfactory result. The moisture part of the model is verified by checking the mass balances in steady state.

$$\frac{dm_3}{dt} = -\beta_{RH} \cdot p_s(T_3) \cdot (\frac{m_3}{m_{3max}(T_3)} - RH_4)$$

$$\frac{dm_4}{dt} = +\beta_{RH} \cdot p_s(T_3) \cdot (\frac{m_3}{m_{3max}(T_3)} - RH_4) - \frac{\delta \cdot p_s(T_4) \cdot (RH_4 - RH_5)}{\mu d} - \frac{\delta \cdot RH_4 \cdot dp_s(T_4)(T_4 - T_5)}{\mu d}$$

$$\frac{dm_5}{dt} = +\beta_{RH} \cdot p_s(T_5) \cdot (\frac{m_6}{m_{3max}(T_6)} - RH_5) + \frac{\delta \cdot p_s(T_4) \cdot (RH_4 - RH_5)}{\mu d} + \frac{\delta \cdot RH_4 \cdot dp_s(T_4)(T_4 - T_5)}{\mu d}$$

$$\frac{dm_6}{dt} = -\beta_{RH} \cdot p_s(T_5) \cdot (\frac{m_6}{m_{3max}(T_6)} - RH_5)$$

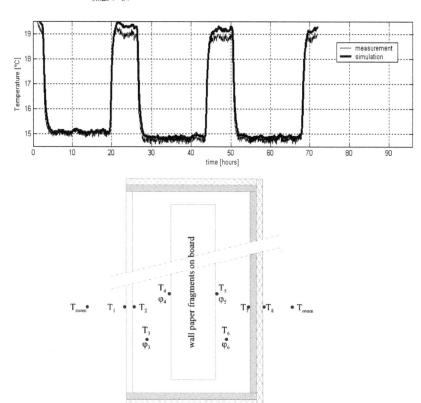

Figure 6.4 Top: The moisture related model equations. Middle: The measured and simulated air temperature in the showcase. Bottom: Schematic view of the modelled quantities (φ = RH)

6.4 SIMULATION RESULTS OF NEW DESIGNS

The models of the previous Section are integrated into a single model in HAMLab. This model is used as a tool to simulate the design options mentioned in Section 6.1.

6.4.1 A new HVAC controller strategy without showcase

As mentioned before, it is not allowed to install new hardware ((de)-humidification) to the current HVAC system. Only modifications of the control strategy are possible. The current HVAC system control strategy is partly based on a feed forward control of the external temperature. A feed back control of the T_a, using a set point of 18 °C, is suggested as possible improvement of control strategy. Figure 6.5 shows the new control model and the simulation results.

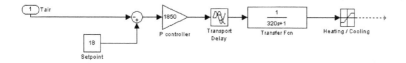

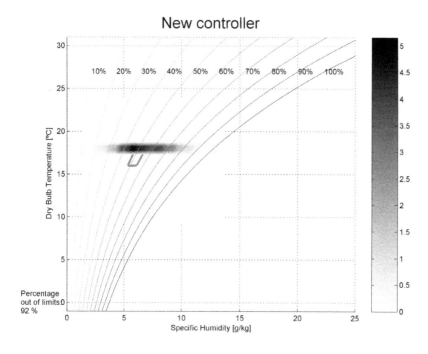

Figure 6.5 Top: The new control model. Bottom: Mollier frequency plot of the simulated climate in the room (see figure 6.2 for explanation. Note: scales are different)

Although the temperature is within the limitations of the control strategy, it is clear that the indoor climate still does not satisfy the recommendations. First, from the Mollier frequency plot, it is concluded that 92% of the measured values are outside the recommended area. Second, from the variations plots, it is concluded that only the T_a changes are within the recommendations now. Third, the RH change per day is still out of limits. The problem we are facing now, is caused by the high moisture gains due to a high number of people inside the room. Without adding (de)-humidification sections to the HVAC system, this problem cannot be solved. As mentioned before such modifications are not allowed and therefore we proceed with the next design option.

6.4.2 The current HVAC system with a showcase

The previous results show that the use of a showcase is almost inevitable. Figure 6.6 provides the simulations results of the climate in a showcase if the showcase model is subjected to the current climate conditions of room.

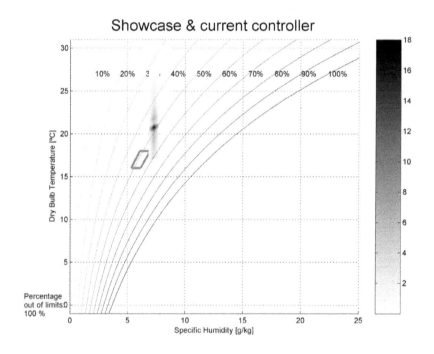

Figure 6.6 Mollier frequency plot of the simulated climate in the showcase (see figure 6.2 for explanation. Note: scales are different)

The indoor climate still does not satisfy the recommendations. This is mainly caused by the large range of indoor air temperature. Also the RH change per day is still out of limits. The problem is still not solved.

6.4.3 A new HVAC controller with a showcase

Figure 6.7 shows the simulated climate in the showcase if the control strategy of Section 6.4.1 is used.

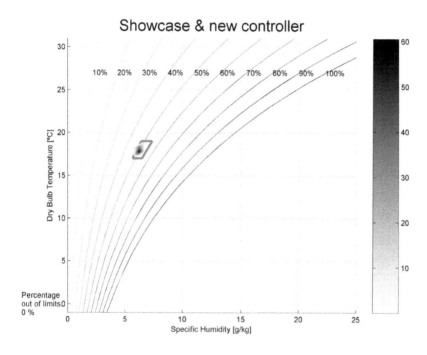

Figure 6.7 Mollier frequency plot of the simulated climate in the showcase (see figure 6.2 for explanation. Note: scales are different)

The indoor climate almost satisfies the recommendations. If the set point is lowered from 18 to 17.5 °C, a perfect climate arises in the showcase for the preservation of wallpaper.

6.5 CONCLUSIONS

If we revisit the key questions mentioned in the introduction, we come to the next conclusions:

1. Recommendations for the climate conditions concerning the preservation of (wall) paper are given.

2. An integrated indoor climate, HVAC system and showcase HAM model has been developed and validated. It is used to simulate the climate conditions for the different design options.

2. It is not possible to satisfy the indoor climate within the recommended limits, exclusively by the use of a new control strategy. Either (de-)humidification sections are needed in the HVAC system, which is not an option here, or the wallpaper fragments have to be placed in a showcase.

3. In order to meet the recommendations, the wallpaper fragments should be placed in a showcase and a similar control strategy as presented in this paper, has to be implemented in order to limit the room air temperature change.

Acknowledgement

The contribution of Anke Wijfelaars, Kim van Zundert and Marcel van Aarle to this work is greatly acknowledged by the authors. Furthermore, the Netherlands Institute for Cultural Heritage is acknowledged for providing measured data concerning the showcase validation.

REFERENCES

Appelbaum, B., 1991. Guide to environmental protection of collections, Madison, CT: Sound View Press.

Henne, E., 1995. Luftbefeuchtung, ISBN: 3486262890

IEA Annex 41, 2005. http://www.ecbcs.org/annexes/annex41.htm

Jutte, B.A.G.H., 1994. Passive conservation (Dutch), Netherlands Institute for Cultural Heritage, internal report

LCM Foundation, 2002. Syllabus for a short course on preventive conservation (Dutch)

Johnson, E.V. & Horgan, J.C., 1979. Protection of the cultural hertitage; technical handbooks for museums and monuments 2. Unesco.

Wijffelaars, J.L., Zundert, K., 2004. Investigation on the indoor climate and wallpaper of a museum. (Dutch) Master Thesis 04.32.W Technische Universiteit Eindhoven

Chapter 7

Optimal set point operation of the climate control of a monumental church

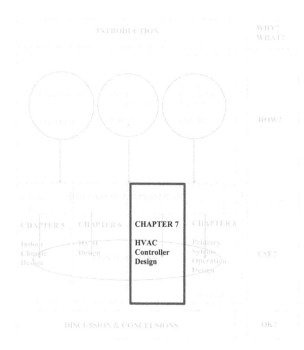

This Chapter presents a case study on the optimal operation of the climate control of the Walloon Church in Delft (Netherlands). It provides a description of constraints for the indoor climate, giving criteria for the indoor air temperature and relative humidity with the focus on the preservation of the monumental organ. The set point operation of the HVAC system is evaluated by simulation using MatLab, ComSol and SimuLink models. The next main model components are presented and combined in a single integrated SimuLink model: 1) a HAMBase SimuLink building model for simulating the indoor temperature and relative humidity, 2) a ComSol PDE model for simulating in a detailed way the dynamic moisture transport in the monumental wood (organ) and 3) a

SimuLink controller model. The building model is validated with measurements. The main advantage of the integrated model is that it directly simulates the impact of HVAC control set point strategies on the indoor climate and the organ. Two types of control strategies are discussed. The first type is a limited indoor air temperature change rate. The second type is a limited indoor air relative humidity change rate. Recommendations from international literature suggest that 1) a change rate of 2 K/h will preserve the interior of churches and 2) a limited drying rate is important for the conservation of monumental wood. This preliminary study shows that a limitation of indoor air temperature change rate of 2 K/h can reduce the peak drying rates by a factor 20 and a limitation of the relative humidity change rate of 2 %/h can reduce the peak drying rates by a factor 50. The second strategy has the disadvantage that the heating time is not constant.

(A.W.M. van Schijndel, H.L. Schellen, D. Limpens–Neilen & M.A.P. van Aarle, published in Proc. of 2TH Int. Conf. On research in building physics, 2003 pp777-784)

7.1 INTRODUCTION

In the Walloon Church in Delft a monumental organ is present which has been restored in the spring of 2000. To prevent damage to the organ again, the indoor climate has to meet certain requirements. Recent studies [Neilen 2003], [Schellen 2002], and [Stappers 2000] have been performed for the preservation of the monumental organ. As a result several adjustments have been made to the heating system. Afterwards, measurements showed that the indoor climate did meet the requirements for preservation of the organ. However, the Walloon Church is not only used for services, but also for several other activities e.g. organ recitals. Since people are sitting in the church without wearing their overcoat, a temperature of 18 to 20 °C is desirable. The result of this rather high temperature for monumental churches, is that the Relative Humidity (RH) of the indoor air may become very low (30%). Since such a low RH can cause damage to the organ, the heating system is restricted. As soon as the RH of the indoor air threatens to drop below 40%, the heating system is shut down. As a result of this restriction it is not possible to reach an indoor temperature of 18 °C in winter when it is freezing outside. Humidification of the indoor air was seen as a possible solution. Due to this measure, the RH of the indoor air remains high enough for preserving the organ and at the same time the indoor air can be heated to the required comfort temperature of 18 °C. As a consequence of humidification during winter there is a risk for condensation and fungal growth on cold surfaces. For that reason a request for further research by simulations was received from the church council. With the help of these simulations an assessment can be made of the potential risks. The main task is to protect the wooden monumental organ from drying induced stresses. In recent studies it is concluded that:

- the increase of drying rate causes a non uniform distribution of the moisture content in dried material and this involves drying induced stress [Kowalski 1999],

- fracture is more likely if the dried body is thick and/or the drying rate is high [Kowalski 2002].

These studies show that the peak drying rate has to be minimized in order to minimize the risk of drying induced stress and fracture.

The main objectives are:

- Development of a single model for simulating the indoor climate, the moisture distribution in the wood of the organ and the HVAC system.

- Evaluation of the current set point operation strategy of the HVAC system in the Walloon church.

- Development and evaluation of new strategies including RH control

7.2 MODELING

7.2.1 The church indoor climate model using HAMBase SimuLink

The indoor climate us simulated using HAMBase (see PART I).

Details of the geometry, material properties and boundary conditions of the HAMBase church model can be found in [Schellen 2002] and [Neilen 2003]. In SimuLink, the HAMBase model is visualized by a single block with continuous input and output connections. The input signal of the HAMBase SimuLink model is a vector containing for each zone heating/cooling power and moisture sources/sinks. The output signal contains for each zone the comfort temperature, the air temperature and RH. In figure 7.1 the input/output structure for the church model (containing 2 zones: church and attic) is shown:

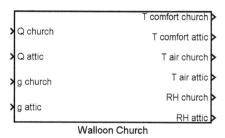

Figure 7.1 The church model in SimuLink. With Q = heating power [W], g =
humidification [kg/s], T comfort = comfort temperature [°C], T air = air temperature
[°C], RH air = air relative humidity [-]

7.2.2 The moisture transport model using ComSol
See Section 3.2

Although ComSol is well equipped for solving complex building physics problems (see
Chapter 3), the ComSol model in this paper is quit simple. The emphasis of this study is
not on the complexity of the individual models but on the complexity of the
combination of models. The moisture transport is assumed to be 1D and dominated by
vapour transport. This means for the PDE coefficients of (7.1a –7. 1c) :

$$u = w \qquad (7.1a)$$
$$d_a = 1 \qquad (7.1b)$$
$$c = D_w \qquad (7.1c)$$
$$g = \beta_{RH} \cdot (RH - RH_{surface}(w)) \qquad (7.1d)$$
$$\alpha = \beta = \gamma = a = f = q = h = r = 0 \qquad (7.1e)$$

where w= moisture content [kg/m³], D_w= moisture diffusivity [m²/s], β_{RH} = surface
coefficient of vapour transfer [kg/m²s], RH= indoor air relative humidity, $RH_{surface}$ (w) =
relative humidity at surface calculated from the hygroscopic curve. Furthermore it is

assumed that the diffusion coefficient is constant and the moisture retention curve is linear in the range of 20% < RH < 90%. The thickness of the wood is 1 cm.

The ComSol model is exported to SimuLink. In figure 7.2 the SimuLink model and its input/output structure is shown:

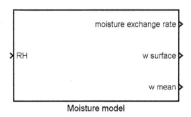

Figure 7.2 The moisture transport model in SimuLink with the moisture exchange rate in [kg/m²s], $w_{surface}$ = moisture content near the wood surface [kg/m³] and w_{mean} = mean moisture content of the wood [kg/m³].

The implementation of a more accurate model for the moisture transport and moisture induced stresses is left for future research.

7.2.3 The controller (Proportional) using SimuLink

In figure 7.3 the controller model is shown:

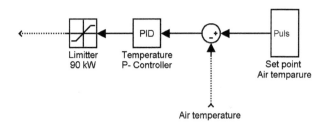

Figure 7.3 The controller model.

The set point of the air temperature is generated by a pulse block with properties: Period: 1 week, start time 04.00 o'clock Sunday, duration: 12 hours, lower value: 10 °C higher value 20 °C. The input of the PID controller consists of the set point minus the actual air temperature. The settings of the PID controller are: P = 107, I=D=0, so in this case it acts like a proportional controller. The output of the controller is limited between 0 en 90 kW.

7.2.4 The complete model in SimuLink

The complete model consists of the models of the church, wood and controller. In figure 7.4 the complete model is shown:

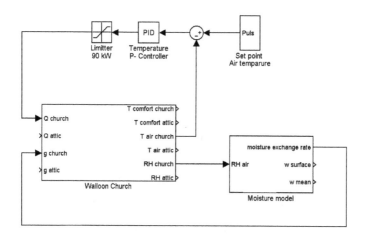

Figure 7.4 The complete model.

There are two closed circuits:

a) an output of the church, the air temperature, is connected to input of the controller and the output of the controller, heating power, is connected to an input (heating/cooling) of the church.

b) another output of the church, relative humidity, is connected to the input of the wood and an output of the wood, moisture exchange rate is connected to an input (humidification) of the church.

In this case, the amount of wood compared with other hygroscopic material (walls) is small. So the influence of the moisture exchange rate of wood on the indoor climate is also small. The connection between the output of the wood and input of the church therefore may be omitted.

7.3 RESULTS

7.3.1 Validation of the HAMBase model

The HAMBase model has been subjected to a validation study by [Neilen 2003]. The measured and simulated air temperature and relative humidity of one month (December 2000) are compared. In figure 7.5 the results are shown:

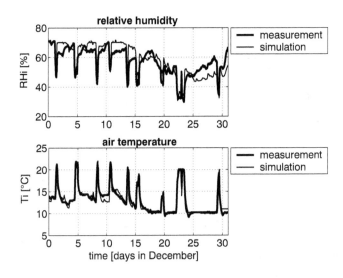

Figure 7.5 Validation of the HAMBase model. The measured and simulated air temperature and relative humidity during one month (December 2000) are compared [Neilen 2003].

Figure 7.5 shows an acceptable agreement between simulated and measured results.

7.3.2 Validation of the ComSol model

In Section 3.3.1 a 1D-moisture transport model in ComSol is validated with satisfactory result. The same model is used, but with other material properties (wood) in [Schellen 2002]. [Schellen 2002] studied the drying and wetting of wood by a fluctuating air relative humidity (35% < RH < 85%). In figure 7.6 the drying result is shown:

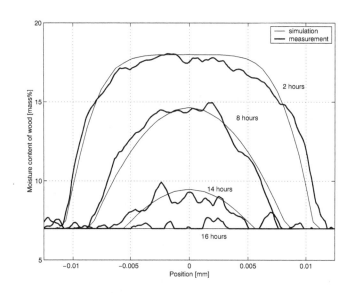

Figure 7.6 Simulation and measurement of moisture profiles in case of drying of a cylinder of wood (diameter 25 mm) by a step in relative humidity from 85% to 35% [Schellen 2002].

Apparently the model can be used for simulating the drying of wood.

7.3.3 Drying rates

The moisture content near the surface and the drying rate (= rate of change of moisture content near the surface) is studied, by using the model of figure 7.4, for 2 cases: a) no heating and b) full heating capacity, i.e. no limitations in air temperature or relative humidity change rate. The simulation period is again one month (December 2000). For all following case studies, the set point operation of figure 7.3 is used. This means that the church is heated 4 times a month. In figure 7.7, the indoor air temperature, the relative humidity and the moisture content of the wood near the surface is shown for the 2 cases, no heating and maximum capacity:

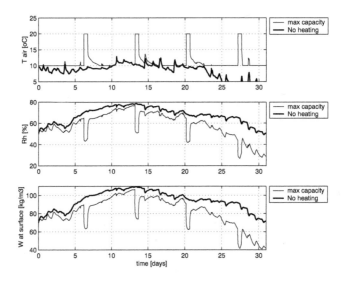

Figure 7.7 The indoor air temperature, the relative humidity and the moisture content of the wood near the surface. The simulation period is December 2000.

In figure 7.8, the drying rate is shown during a period of 1 day (starting Saturday 0.00 o'clock) for the 2 cases, no heating and maximum capacity:

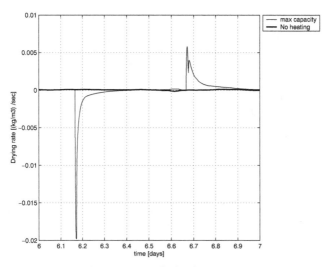

Figure 7.8 The drying rate during a period of 1 day (starting Saturday 0.00 o'clock).

A negative drying rate indicates that water is transported from the material to the surroundings i.e. drying. In figure 7.9, the peak-drying rate defined as the absolute value of the drying rate for that same period is shown on a log scale

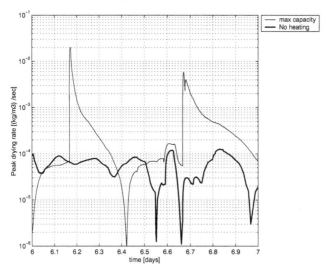

Figure 7.9. The peak drying rate during a period of 1 day (starting Saturday 0.00 o'clock).

From figure 7.9 the difference in peak drying rate for the case of no heating and the case of full heating capacity is very clear: The peak drying rate, in the case of full heating, is of order ~100 times larger than in case of no heating. These peaks can cause drying induced stresses [Kowalski 1999] and have to be minimized to prevent possible damaging of the wood. In the next Section some alternative set point operations will be discussed.

7.4 SET POINT OPERATION STUDY

In Section 7.3 the results of the control strategies: No heating and full heating capacity are already evaluated. In this Section two alternative set point operations will be discussed.

7.4.1 Limitation of the air temperature change rate

Recommendations from international literature [Schellen 2002] suggest that an air temperature change rate of 2 K/h will preserve the interior of churches. The limitation of the air temperature change rate is modelled by a 'Rate Limiter' block of SimuLink. The complete model including the set point temperature Rate Limiter is shown in figure 7.10:

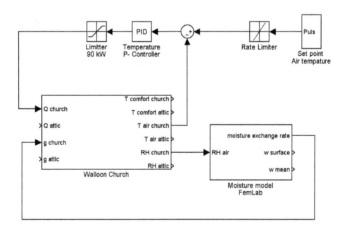

Figure 7.10 The complete model including a 'Rate Limiter' block for the limitation of the air temperature change rate.

The output of the air temperature set point is connected to a 'Rate Limiter' block of SimuLink. This block has two parameters: the Rising slew rate (R) and the falling slew rate (F). This block is modelled by:

$$rate = \frac{u(i) - y(i-1)}{t(i) - t(i-1)}, \qquad (7.2a)$$

$$if\ rate > R : y(i) = \Delta t \cdot R + y(i-1) \quad (7.2b)$$

$$if\ rate < F : y(i) = \Delta t \cdot F + y(i-1) \quad (7.2c)$$

$$else : y(i) = u(i) \qquad (7.2d)$$

where u = input of the block, y = output of the block, t = time, Δt=t(i)-t(i-1), (i) = actual time step, (i-1) = previous time step. In figure 7.11, the indoor air temperature, the relative humidity and the moisture content of the wood near the surface during a period of 1 day (starting Saturday 0.00 o'clock) is shown, for different cases including limitation of the air temperature heating change rate of 1.5 K/h and 2.5 K/h. (This means for the parameters of the rate limiter that the rising slew rate R equals 1.5/3600 resp. 2.5/3600 and the falling slew rate F equals -8 in both cases).

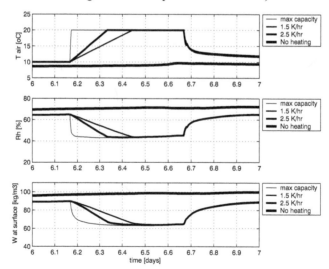

Figure 7.11 The indoor air temperature, the relative humidity and the moisture content of the wood near the surface during a period of 1 day (starting Saturday 0.00 o'clock).

In figure 7.12 the peak-drying rate is shown for these cases:

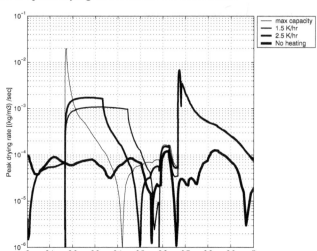

Figure 7.12 The peak drying rate during a period of 1 day (starting Saturday 0.00 o'clock).

From figure 7.12 it follows that a limitation of the temperature change rate of 2 K/h reduces the peak drying rates by an order of ~10 compared with no limited temperature heating change. However, the peak factor is still an order of ~10 higher compared with no heating.

7.4.2 Limitation of the relative humidity change rate
A more challenging task is to model the heating of the church with a limitation of the relative humidity. In figure 7.13 the complete model including the relative humidity change limiter is shown.

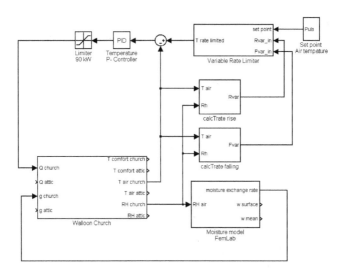

Figure 7.13 The complete model including the relative humidity change limiter.

The output of the air temperature set point block is connected to a new developed 'Variable Rate Limiter' block. This block has 3 inputs 1) the air temperature set point, 2) the (variable) rising slew rate (R_{var}) and the (variable) falling slew rate (F_{var}). The output of this block is analog to the 'Rate Limiter'. It can now also handle variable rising and falling slew rates. The slew rates R_{var} and F_{var} are calculated by the block 'calcTrate'. The inputs of this block are the air temperature and the relative humidity. The output consists of the slew rate of the air temperature. The output is calculated from standard psychometrics functions, already programmed in MatLab and the only parameter of this block, ΔRH, the relative humidity change rate in %/h:

$$T_{rate} = \left| T_a - T_{dew} \left(\frac{RH \cdot p_s(T_a)}{RH - \Delta RH} \right) \right| / 3600 \quad (7.3)$$

where T_{rate} = computed temperature change rate [°C/sec], T_a = air temperature [°C], T_{dew} = dewpoint function [°C], RH = relative humidity [%], p_s = saturation pressure function [Pa], ΔRh = maximum relative humidity change rate [%/h].

121

In figure 7.14, the indoor air temperature, the relative humidity and the moisture content of the wood near the surface during a period of 1 day (starting Saturday 0.00 o'clock), is shown for different cases including limitation of the relative humidity change rate of 2%/h and 5%/h:

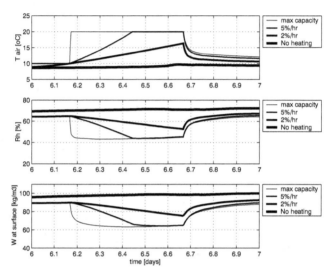

Figure 7.14 The indoor air temperature, the relative humidity and the moisture content of the wood near the surface during a period of 1 day (starting Saturday 0.00 o'clock

In figure 7.15 the peak-drying rate is shown for these cases:

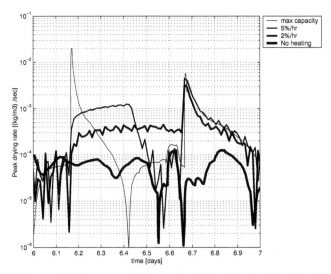

Figure 7.15 The peak drying rate during a period of 1 day (starting Saturday 0.00 o'clock)

From figure 7.15 it follows that a limitation of the relative humidity change rate of 2%/h reduces the peak drying rates by an order of ~50 compared with no limited temperature heating change. However, the peak factor is still an order of ~5 higher compared with no heating. Notice that in this case, the maximum occurring indoor air temperature is 16 °C. This is far below the set point temperature of 20 °C.

7.5 DISCUSSION

7.5.1 Comparing the control strategies
The control strategies of limiting the temperature or relative humidity change rates look rather familiar for the peak drying rates (see figure 7.12 and 7.15). However, they are not the same. The difference in controlling strategy is shown in figure 7.16. In this figure the air temperatures of figures 7.11 and 7.14 are combined and shown for 2 different days.

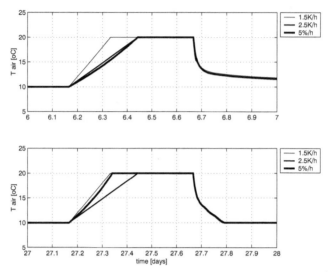

Figure 7.16 The air temperature for the controlling strategies. The upper part shows the air temperature at 6 December, the lower part shows the air temperature at 27 December.

From figure 7.16 it follows that the 5%/h relative humidity rate limitation approaches the 2.5 K/h temperature rate at 6 December, but it approaches the 1.5 K/h temperature rate at 27 December. Also from figure 7.16 it can be seen that the time needed for heating the church to 20 °C in case of the 5%/h relative humidity rate limitation varies from 0.25 days (= 6 hours) on 6 December to 0.15 days (=3.6 hours) on 27 December. These differences show that the 2 strategies, temperature and relative humidity change rate limitation are quit different. This is also shown in figure 7.17, where the air temperature (variable) Rising slew rate (R_{var}) is plotted against time for the relative humidity change rate limitations.

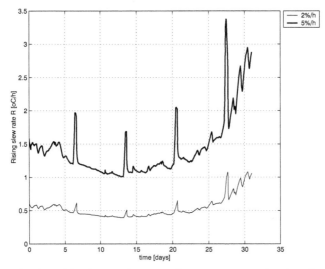

Figure 7.17 The (variable) Rising slew rate (R_{var}) versus time for the relative humidity change rate limitations. The time period is December 2000.

From figure 7.17 it follows that the air temperature Rising slew rate (Rvar) can change by a factor ~3 during the time period in case of relative humidity change rate limitation.

7.5.2 Optimal set point operation

All heating strategies presented in this paper are now evaluated:

- The best solution to prevent high peak drying rates is no heating. Due to thermal discomfort, this is not an acceptable solution.

- The worst solution to prevent high peak drying rates is full heating capacity. From figure 7.9 it follows that the peak drying rate is of order ~100 times larger than in case of no heating. This is seen as the main cause of the damaging of the previous organ of the Walloon church [Schellen 2002], [Neilen 2003] and is therefore not acceptable.

- Two possibilities to limit the peak drying rates are studied: Limitation of the change rate of the air temperature and the relative humidity. Both are rather familiar in case of the limitation of the peak drying rates. The disadvantages of a limitation of the relative humidity change rate compared to a limitation of the air temperature change rate are:

- The time to heat the church is not constant.

- A more complex controller is needed (compare figure 7.10 with 7.13).

Therefore a limitation of the air temperature change rate is preferred.

The studied parameters of the limitation of the air temperature change rates of 1.5 and 2.5 K/h are based on a recent study of [Schellen 2002]. In this study it is suggested that a limitation of the indoor air temperature change rate of 2 K/h will preserve the interior of churches. A more detailed study based on drying induced stresses in wood is needed to further investigate this parameter and is left for future research.

7.6 CONCLUSIONS

The set point operation of the HVAC system is evaluated by simulation using MatLab, ComSol and SimuLink models. The following main model components are presented and combined in a single integrated SimuLink model: 1) a HAMBase SimuLink building model for simulating the indoor temperature and relative humidity, 2) a ComSol PDE model for simulating detailed dynamic moisture transport in the monumental wood (organ) and 3) a SimuLink controller model. The main advantage of the integrated model is that it directly simulates the impact of HVAC control set point strategies on the indoor climate and the organ in terms of peak drying rates.

Two types of control strategies are discussed. The first type is a limited indoor air temperature change rate. The second type is a limited indoor air relative humidity change rate. A limitation of the air temperature change rate of 1.5 to 2.5 K/h is preferred. A more detailed study based on drying induced stress in wood is needed to further refine the temperature change rate

REFERENCES

COMSOL AB, 2000, ComSol Version 2.0 pre, Reference Manual

Kowalski, S.J. & Rybicki, A., 1999, Computer Simulation of Drying Optimal Control, transport in Porous Media vol34 pp227-238.

Kowalski S.J., 2002, Modeling of fracture phenomena in dried materials. Chemical Engineering Journal vol86 pp145-151.

Neilen, D., Schellen, H.L.& Aarle, M.A.P. van, 2003, Characterizing and comparing monumental churches and their heating performance; Sec. Int. Building Physics conference Leuven.

Schellen, H.L., 2002, Heating Monumental Churches, Indoor Climate and Preservation of Cultural heritage; PhD Dissertation, Eindhoven University of Technology.

Stappers, M.H.L., 2000, De Waalse Kerk in Delft; onderzoek naar het behoud van een monumentaal orgel Master thesis FAGO 00.03.W. Eindhoven University of Technology.

Chapter 8

Optimal operation of a hospital power plant

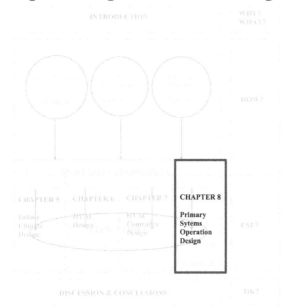

This chapter presents a case study on the optimal operation of a complex hospital installation. A mathematical model has been developed which is based on energy balances of the installed components. Manufacturer specifications of the components are used for calculating parameters. The model is formulated using vector equations. Advantages of this type of model formulations are presented. The tools used for optimization are a custom developed back tracking method for calculating a good starting point and a SQP optimization tool for finding the optimum. Detailed control strategies are calculated for 3 types of optimization strategies simulated. The simulation results show the impact of the choice of a control strategy on the optimized operation. The results are also applicable for on line set point optimization.

(A.W.M. van Schijndel, published in Energy and Buildings, 2002, vol34/10 pp1061-1071

8.1 INTRODUCTION

It is no longer sufficient to design a system which performs the desired task while observing constraints imposed by desired safety, economic output and other considerations. Due to the need to increase efficiency, it has become essential to optimize the process to minimize or maximize a chosen variable. This variable is known as the objective function (or optimization criteria). Profit or cost is often used as objective function, though many other aspects, such as efficiency, energy, weight, size, etc., may also be optimized depending on the particular application. For example, a central heating plant can be designed for a desired amount of heat production. If a specific system is chosen, the various parameters such as operating temperature, heat exchange area, water flow, etc. may be selected over a wide range of specifications [Jaluria 1998]. All the designs may be acceptable because they satisfy the given requirements and constraints. It may however be necessary to seek an optimal design that will, for instance, consume the least amount of energy and is the most cost effective. An important optimization criterion of power plants is the efficiency of generating electricity. A small increase in efficiency can give a large increase in the profit. Nowadays a lot of power plants use the produced heat for heating purposes like the power plant at the academic hospital Groningen. In this case of combined heat and power (CHP), the efficiency of generating electricity is not the only optimization criteria because the produced heat can also reduce cost. Also present in the installation, is an absorption chiller, which generates cooling from heat. So in the summer, when cooling is needed, the produced heat of the power plant is not wasted but can be used for cooling purposes. Energy saving criteria are very important in this case because the CHP part of the power plant is partly funded by the Dutch government. The main reasons for subsidizing CHP's are energy savings and CO_2 reduction. Because of its complexity, it is very difficult to design an optimal operating strategy for the power plant. A computer model may do that.

A recent similar study on optimal operation has been done by [Dentice 2001]. The subject is also a complex thermal plant that is quite similar to the one presented in this paper. Although both models are based on the same principles, the model syntaxes are very different. Benefits of using the model syntax in this paper are presented in Section 8.3. Both studies use similar optimization solving techniques: Start with the best point from a selection of trial points and refine the optimum. An important extension of the work of [Dentice 2001] is that in this paper the plant analyses and optimizations are performed on a yearly base. The papers of [Fu 2000] and [Ahn 2001] deal with more complex modeling of components. The thermal and electric efficiencies in [Fu 2000] en [Ahn 2001] are dependent on temperatures. The thermal and electric efficiencies presented in this paper are modeled more low-level and are dependent on engine percentage loads. In [Krause et al. 1999] the electrical efficiency of a CHP is also successfully modeled by engine percentage loads. Other studies of optimal operation [Kruse et al. 1999], [Bojic et al. 2000] and [Benonysson 1995] deal with district heating systems using CHP's.

For modeling and optimization purposes a system is needed which generates an optimized output from certain input. This system contains a model and an optimization routine. The inputs of the model can be divided into a non-controllable input i(t) and a controllable input c(t). The output of the model is o(t). The output of the model is used as input for the controller (optimization routine). The controller minimizes the impact of the non-controllable input i(t) by manipulating c(t). Figure 8.1 shows the principle of modeling and optimization.

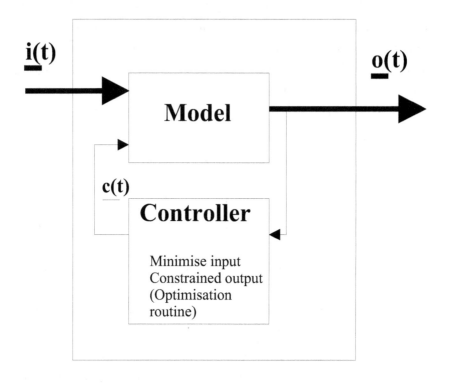

Figure 8.1 The relation between model and optimization.

In order to develop an optimization program seven steps are used:

1) Design a model

2) Define non-controllable and controllable inputs and the output

3) Define constraints

4) Define optimization criteria

5) Build a numerical model

6) Select an appropriate time scale

7) Build a numerical optimization routine and calculate the optima

8.2 THE APPLICATION

The central utility plant of the Academic Hospital Groningen has been selected as the subject for modeling and optimization. The installation meets the energy needs of the hospital. The installation is designed for producing domestic hot water and heating (30 MW), steam (8 MW), cooling (12 MW) and electricity (7.5 MW). Typical energy aspects of the installation, built in 1995, are: total heat consumption including the absorption chiller: 66 GWh; total electricity consumption: 27 GWh; gas consumption of the gas engines: 13,000,000 m³; and gas consumption of the boilers: 3,000,000 m³. The central power plant consists of 3 boilers, each with a capacity of 12 MW, 5 gas engines, each generating 2 MW electricity and 2.5 MW heat, 1 absorption chiller for generating 3 MW cooling from heat. Distributed over the hospital terrain, 7 cooling units are connected to the cooling circuit, each contains a mechanical chiller of 1.1 MW and an ice storage tank (TES) of 16 GJ. Figure 8.2 shows the central power plant in detail

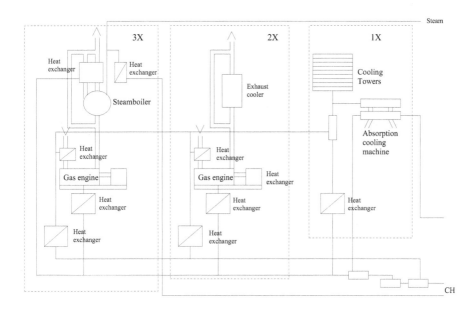

Figure 8.2 The power plant (the 7 cooling units, connected with the absorption chiller are not shown in this figure).

8.3 MODELING AND OPTIMIZATION

In this Section the steps used for optimization, are described. The model equations are based on a quasi-steady approach (the variables are approximate constant between two time steps) and contain scalars (normal letter type) and vector variables (bold letter) which contain values for this variable at each time step for a whole year. Compared with the model syntax of [Dentice 2001], this syntax has the following benefits:

1) Time efficient computation: The vector-based model equations are identical to the equations programmed in MatLab. Because MatLab is a vector oriented numerical program, it is very time efficient on vector operations and optimization problems.

2) Model validation: By putting the model equations, exactly as they are presented in this paper, into a symbolic computation software program like Maple; the model equations can be theoretically validated.

In the project this benefit has successfully been used to correct preliminary errors in the model equations. The presented model syntax has also a drawback: It is perhaps more difficult to follow than a common model syntax.

8.3.1 Design a model

The model is based on energy flows. In Figure 8.3 the energy flows (heat, cooling, electricity and primary energy) are shown.

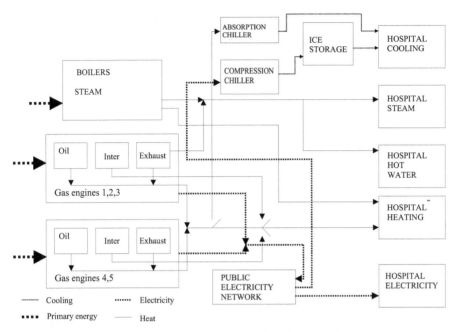

Figure 8.3 The energy flows of the power plant and cooling installations

8.3.2 Define non-controllable and controllable inputs and the output

The non controllable input consists of: The demand for cooling (**QcA**), heating (**QhA**), electricity (**ElA**), steam (**msA**), hot water (**mlA**) and the energy price for gas (**gpB** (boilers), **gpG** (gas engines)) and electricity (**Epb, Eps**). The values of these variables are based on measurements and available utility rates. The controllable input consists of: The set points of the gas engine group 1 (**G13**) and group 2 (**G45**) and the set points of the mechanical chiller (**CCD**) and the absorption chiller (**ACD**). The output consists of: Total profit (**Totp**), the primary energy (**Qprim**), the wasted useful energy (**Ekill**) and the cooling energy stored in the ice storage tanks (**QcB**).

8.3.3 Define constraints

The controllable inputs (setpoints of: **G13**, **G45**, **CCD**, **ACD**) are constrained by a minimum of 0 (represents switching off) and a maximum of 1 (represents maximum capacity). The state-of-charge of the ice storage tanks (**QcBsum**) is constrained with 0 (represents empty tanks) and QcB_{max} (full loaded tanks). Furthermore, the academic hospital is connected to the public electricity grid. The supply of electricity is therefore guaranteed. The capacity of the boilers is much higher than the maximum heat demand, the supply of heat is therefore guaranteed. So no extra constraints are needed for electricity and heating. Figure 8.4 shows the principle of modeling and optimization together with the non-controllable and controllable inputs and the output and constraints.

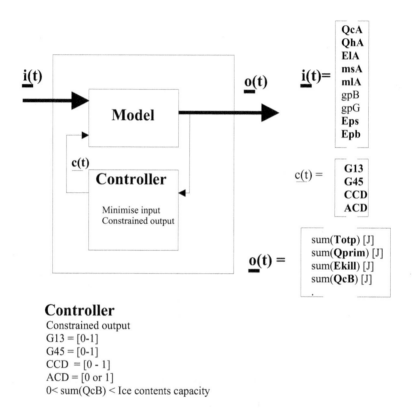

Controller
Constrained output
G13 = [0-1]
G45 = [0-1]
CCD = [0 - 1]
ACD = [0 or 1]
0< sum(QcB) < Ice contents capacity

Figure 8.4 The principle of modeling and optimization.

8.3.4 Define optimization criteria

The goal is to optimize the operation of the controllable input variables (set points) of: gas engine group 1 (**G13**), group 2 (**G45**), the compressor (**CCD**) and absorption chillers (**ACD**), using an optimization strategy to minimize the chosen cost function. The next three optimization strategies are considered:

Strategy 1: Maximize the profit,

Strategy 2: Minimization of the primary energy use with the constraint that the profit is greater than zero,

Strategy 3: Minimize the primary energy use.

8.3.5 Build a numerical model

In Table 5.I the input and output of the present components of the installation are summarized.

Table 5.I Input and output descriptions of the components

Description of component	Input	Output
Hospital cooling	Cooling needs	-
Hospital steam	Steam needs	-
Hospital hot water	Hot water needs	-
Hospital heat	Heating needs	-
Hospital electricity	Electricity needs	-
Gas engines 1,2,3	primary energy (gas)	electricity, heat
Gas engines 4, 5	primary energy (gas)	electricity, heat
Boilers	primary energy (gas)	heat
Absorption chiller (ACD)	heat	cooling
Compression chiller (CCD)	electricity	cooling
Ice storage	cooling	cooling
Public electricity network	electricity	electricity
Switch ACD	switching of the ACD when heat is available	
Switch inter coolers	switching of inter coolers for extra heating needs	

The 5 gas engines are divided into 2 groups. First **G13** are the gas engines 1, 2 and 3 with sub components: 1) electricity, 2) exhaust-heat exchanger to steam, 3) exhaust-, inter- and oil-heat exchanger for heating the academic hospital. Second **G45** are the gas engines 4 and 5 with sub components: 1) electricity, 2) exhaust-, inter- and oil-heat exchanger for heating the academic hospital. The efficiencies of the gas engine components depend on the operation of the engines themselves. For example, when the intercoolers are disconnected the electricity production efficiency increases and the heat production efficiency decreases. The three boilers, the seven mechanical chillers (**CCD**) and the ice storage tanks are considered each as one system. When more heat is produced than demanded by the hospital, the absorption-chiller (**ACD**) can be switched on. The excessive amount of heat is then used for cooling. The ice storage has only a small heat gain from the surroundings. The efficiencies of the CHP units and chillers depend on their setpoints, according to the manufacture specifications. All other efficiencies are modeled as constants.

The presented model is vector oriented. Furthermore the next functions are used in the model representation (**x** is a vector variable):

$$posvec(\mathbf{x}) = 1/2 * abs(\mathbf{x}) + 1/2 \, \mathbf{x} \qquad (8.1)$$

$$negvec(\mathbf{x}) = -1/2 \, abs(\mathbf{x}) + 1/2 \, \mathbf{x} \qquad (8.2)$$

The next simple property can be derived from (1) and (2):

$$\mathbf{x} = posvec(\mathbf{x}) + negvec(\mathbf{x}) \qquad (8.3)$$

The above functions are used to cope with different parameters for positive and negative elements of x. For example, to calculate the electricity profit, positive elements of the electricity balance vector have to be multiplied by the sale price of electricity and negative elements have to be multiplied by the purchase price of electricity (see Eq 8.20). Another function used in the model is:

$$sign(x): \begin{cases} 1 \text{ if } x > 0 \\ 0 \text{ if } x = 0 \\ -1 \text{ if } x < 0 \end{cases} \qquad (8.4)$$

Because **x** is a vector, the result of sign(**x**) is also a vector with positive values of **x** replaced by 1's, negative values of **x** replaced by -1's and zero values of **x** are unchanged. This function is used for switching the supply heat to the absorption chiller (see Eq 8.10).

model equations

Figure 8.3 is used as a guideline for the model equations below. The primary energy power equations for the gas engines are:

$$P13 = G13 \cdot g_{\max 13} \cdot H \tag{8.5}$$

$$P45 = G45 \cdot g_{\max 45} \cdot H \tag{8.6}$$

The electrical power of the compressor chillers:

$$EIC = CCD \cdot CCD_{\max} \tag{8.7}$$

The electricity balance (negative values mean a shortage of electricity):

$$Elbal = P13 \cdot \eta_{E13} + P45 \cdot \eta_{E45} - EIC - ElA \tag{8.8}$$

The heat balance:

$$QCHbal = P13 \cdot (\eta_{h13A} + \eta_{h13I}) + P45 \cdot (\eta_{h45A} + \eta_{h45I}) - QhA \tag{8.9}$$

The electric and thermal efficiencies of the CHP units depend on their setpoints, according to the manufacture specifications. Switching of the absorption chiller is given by positive elements of **QCHbal** meaning that the absorption chiller is allowed to switch on and is represented by ones in the variable **QhSW**. Negative elements of

QCHbal mean that the absorption chiller is not allowed to switch on and is represented by zeros):

$$\mathbf{QhSW} = posvec(sign(\mathbf{QCHbal})) \tag{8.10}$$

The non-used heat due to the heat supply to the absorption chiller exceeding the maximum design heat supply:

$$\mathbf{QACDback} = posvec(\mathbf{QCHbal} - QcS\max) \tag{8.11}$$

The used heat from the gas engines to the absorption chiller and heat supply for the central heating of the building:

$$\mathbf{QhACD} = \mathbf{QhSW} \cdot \mathbf{ACD} \cdot (posvec(\mathbf{QCHbal}) - \mathbf{QACDback}) \tag{8.12}$$

$$\mathbf{QhCH} = \mathbf{P13} \cdot (\eta_{h13A} + \eta_{h13I}) + \mathbf{P45} \cdot (\eta_{h45A} + \eta_{h45I}) - \mathbf{QhACD} \tag{8.13}$$

The heat balance of the boilers (positive values mean a heat demand for the boilers), the heat demand of the boilers and the gas supply to the boilers:

$$\mathbf{QBbal} = -negvec(\mathbf{QhCH} - \mathbf{QhA}) - \mathbf{P13} \cdot \eta_{h13S} + \mathbf{msA} \cdot (h_{st} - h_{cwa}) + \mathbf{mlA} \cdot (h_{hwa} - h_{cwa}) \tag{8.14}$$

$$\mathbf{QB} = posvec(\mathbf{QBbal}) \tag{8.15}$$

$$\mathbf{mgB} = \frac{\mathbf{QB}}{H \cdot \eta_{hK}} \tag{8.16}$$

The cooling power supply for the ice storage tanks:

$$\mathbf{QcB} = \mathbf{ElC} \cdot \eta_{cC} + negvec(\mathbf{QhACD} \cdot \eta_{cS} - \mathbf{QcA}) \tag{8.17}$$

The thermal efficiencies of the chillers depend on their set points, according to the manufacture specifications. If the cooling supply of the absorption chillers exceeds the cooling demand, the negvec function replaces positive elements by zeros, representing that the absorption chiller cannot charge the ice storage system. The wasted useful heat[*] supplied directly or indirectly by the gas engines is:

$$\textbf{Ekill} = -negvec(\textbf{QBbal}) + posvec(\textbf{QhCH} - \textbf{QhA}) + posvec(\textbf{QhACD} \cdot \eta_{cs} - \textbf{QcA}) \quad (8.18)$$

The three terms at the right hand side of the equation in Eq. 8.18 represent the wasted useful energy for respectively the steam production; the heating of the hospital and cooling produced by the absorption chiller. The total primary energy equivalent of the gas flow and electricity supply:

$$\textbf{Qprim} = H \cdot \textbf{mgB} + \textbf{P13} + \textbf{P45} - negvec(\textbf{Elbal}) / \eta_{Epub} \qquad (8.19)$$

The electricity profit (positive values mean profits), the gas cost (positive values mean costs) and the total profit (positive values mean profits):

$$\textbf{EP} = \textbf{Eps} \cdot posvec(\textbf{Elbal}) + \textbf{Epb} \cdot negvec(\textbf{Elbal}) \qquad (8.20)$$

$$\textbf{gP} = gpB \cdot \textbf{mgB} + gpG \cdot (g_{max\,13} \cdot \textbf{G13} + g_{max\,45} \cdot \textbf{G45}) \qquad (8.21)$$

$$\textbf{Totp} = \textbf{EP} - \textbf{gP} \qquad (8.22)$$

Note that the electricity prices for purchase and sale are vectors because they depend on the time (day and night values). The gas prices for the gas engines and boilers are different fixed prices.

---

[*] Wasted useful heat is the overcapacity of produced heat that cannot be used anymore.

8.3.6 Select an appropriate time scale

The time scale is a very important factor when using the quasi-steady approach. The switching of gas engines is limited to once during the day (16 hours) and once during the night (8 hours). The ice storage can be loaded from empty to full capacity within 8 hours. The electricity prices are different during the day and night. When selecting a time step of 16 hours (day) and 8 hours (night), the optimization horizon can be limited to 24 hours. In this situation it is possible to generate cooling from the compressor chiller to the ice storage during the night when the electricity is cheaper and use it during the day. Using a total simulation period of one year and a resolution of 2 steps a day, each variable contains 730 states.

8.3.7 Build a numerical optimization routine and calculate the optima

The goal is to optimize the controllable inputs in such a way that the output satisfies the given criteria and the constraints. A well-known problem with nonlinear optimization is that the calculated solution is not the global optimum but a local optimum. In order to prevent this and also to get a fast convergence, a good starting point is necessary. A backtracking method [Schijndel 1998] is used for this problem and it consists of the next steps:

- divide all controllable inputs into discrete steps
- calculate all possible combinations of the controllable inputs
- simulate all combinations
- select the best solution (back tracking) from all possible solutions given the optimization criteria.

Using the back tracking method, it is necessary to compute all combinations from the next values for the controllable inputs: **G13** and **G45** (0, 0.25, 0.5, 0.75 ,1), **CCD** : (0, 0.20, 0.35, 0.50 ,1) and **ACD** (0 or 1). These are 250 possibilities per period. A time frame of two periods must be evaluated. Given a day and night period then 62500 possibilities are present. Figure 8.5 shows the results of the back tracking method.

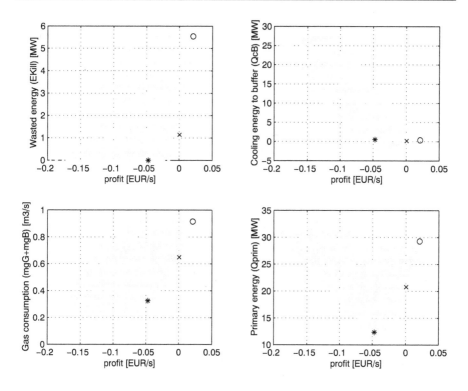

Figure 8.5 The results of the back tracking method. The wasted energy, cooling energy to the buffer, gas consumption and used primary energy against the profit[*].

The wasted energy, cooling energy to the buffer, gas consumption and used primary energy are plotted against the profit. Considering the constraint for the ice buffer, the next optima are selected: The maximum profit (o) , the absolute minimum used primary energy (*) and the minimum used primary energy with a constraint that the profit is positive (+). This method gives a rough estimation of the optima. Because the optimization problem is non-linear and discontinue (see also [Dentice 2001]), a SQP (Sequential Quadratic Programming) routine of the MatLab Optimization Toolbox [Mathwork 1998] is used to refine the optimization.

---

[*] Please note: 0.1 EUR/s = 3.15 million EUR/year; 1MW = 8.8 GWh/year

8.4 RESULTS

8.4.1 The non-controllable input signals of the model i(t)

Figure 8.6 shows some of the non-controllable inputs of the model: Cooling, heating, electricity and steam needed for the academic hospital against time.

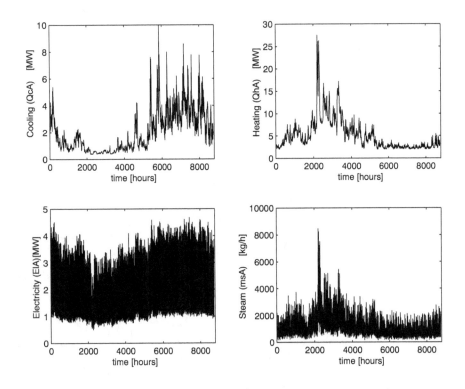

Figure 8.6. The non-controllable inputs of the model: Cooling, heating, electricity and steam needed for the academic hospital against time.

The above time series are simulated time series based on global values. The time period is one year and is shown in hours after 1 October. Other non-controllable inputs are the prices for purchasing and selling electricity and the price of gas.

8.4.2 The optimization results

In Figure 8.7 the output signals: Profit, primary energy used, wasted energy and optimal operation of gas engines 1-3, are shown, when using optimization strategy 2, for a period of one year.

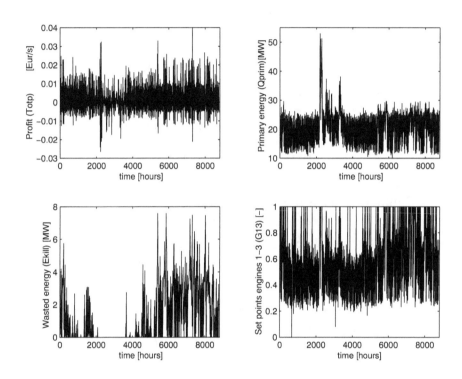

Figure 8.7 The output signals: Profit, used primary energy, wasted energy and set point of gas motors 1-3 against time.

Also details on the optimal operation of gas engines 4-5, mechanical chillers and the absorption chiller are calculated. The yearly characteristics of the 3 strategies are listed in Table 5.II

Table 5.II The year characteristics of the optimization results of the strategies

Year characteristic	Strategy 1	Strategy 2	Strategy 3
Total profit [Eur]	700,000	0	-1,500,000
Gas usage boilers [m3]	1,290,000	1,860,000	4,300,000
Gas usage gas motors [m3]	27,900,000	20,150,000	8,200,000
Total heat demand [MWh]	69,000	69,000	69,000
Total cool demand [MWh]	17,000	17,000	17,000
Electricity for hospital [MWh]	24,000	24,000	24,000
Electricity for cooling [MWh]	800	1,100	4,800
Total electricity produced [MWh]	87,800	62,700	25,500
Total primary energy [MWh]	259,700	195,700	120,000
Wasted useful heat [MWh]	33,700	12,500	0

8.4.3 Comparing the different strategies

When comparing the different strategies, the primary energy versus the profit and the wasted useful heat versus the profit are shown in Figure 8.8.

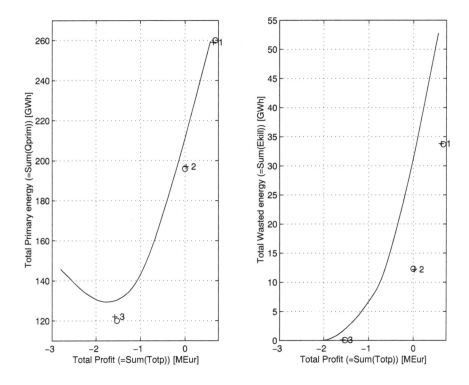

Figure 8.8. The yearly primary energy and the yearly wasted energy versus the yearly profit. Labeling: 'o': result of SQP optimization method; '+': result of the back tracking method; '1,2,3': label for strategy 1,2,3. The line (-) is calculated from varying the set points of all gas engines from 0% to 100% with the assumption that no energy can flow to or from the ice buffer (i.e. assuming no storage capacity of the ice buffer).

The differences between the back tracking method and the SQP method seems to be small on the total scale of possible set points. Still the SQP method gives always a higher profit (at least 0.05 MEur= million Euros) than the back tracking method. When the set points of all the gas engines are held at 1 all year, the presence of the ice buffer can give an extra profit of 0.1 MEur and a reduction of 20 GWh (equal to the amount of heating 1000 average Dutch houses) of the wasted useful heat, each year.

8.4.4 The total efficiency of the power plant

From the simulation results it is possible to look at the overall efficiency of the power plant. In Figure 8.9 part A, the efficiencies of the boilers and gas engines are shown and in part B the fraction of gas consumption of the boilers and gas engines is shown.

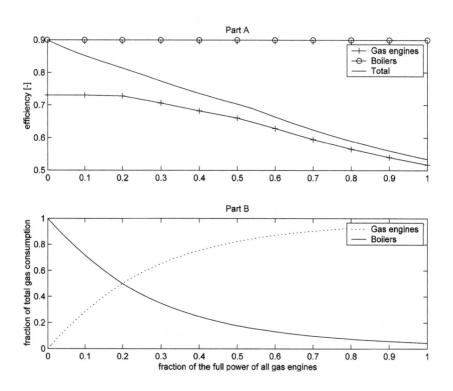

Figure 8.9 Part A, the efficiency of the boilers and gas motors, part B the fraction of gas consumption of the boilers and gas motors.

The total efficiency of the power plant is calculated from the efficiencies of the gas engines and boilers and weighted with the fraction of the total gas consumption of the gas engines and boilers. The gas engines are designed for an overall efficiency of 0.73 and the boilers 0.9. When the gas engines are off all year, the total efficiency equals the efficiency of the boilers. When the mean fraction of full power of the gas engines is 0.20, then both boilers and gas engines have equal gas consumption. The total efficiency is 0.82. When the mean fraction of full power of the gas engines is exceeds 20% the efficiency of the gas engines decreases because the gas engines produce more heat than needed for the hospital. The total efficiency drops to 0.53 when the gas engines are set to 1 all of the year.

8.5 CONCLUSIONS

A model for a complex hospital combined heating, power and cooling plant is presented. The model is based on vector equations. The main advantages of this type of modeling are a time efficient computation and the possibility to check the model analytically. The simulations facilitate an environment for testing different strategies for an optimal operation on a yearly base. Three strategies are used for the optimization of the operation of the power plant: The first optimization strategy is a pure economic one. This strategy gives a profit of 0.7 MEur with 260 GWh of primary energy consumption. This situation can be summarized as a) All gas engines at maximum level during day and night, b) The ice storage tanks filled at night. The second optimization is a pure energy optimization. This strategy gives a loss of 1.5 MEur with only 120 GWh of primary energy consumption. The third optimization is an energy optimization with the constraint that the profit must be positive. This strategy gives a profit of 0.0 MEur with 196 GWh of primary energy consumption. For the last two optimizations, the way the systems should operate is calculated in detail. It is however not possible to summarize these results in a simple way just like the first optimization. The method is also applicable for similar optimization problems. The thermal and electric efficiencies presented in this paper are modeled low-level as being dependent on load fraction only. Implementation of more complex models is left for future research.

REFERENCES

Schijndel, A.W.M. van. 1998. Optimization of the operation of the heating, power and cooling installation of the Academic Hospital Groningen, report no. NR-2029, Eindhoven University of Technology

Jaluria, Y. 1998. Design and optimization of thermal systems, McGraw-Hill

Fu L, Jiang, Y. 2000. Optimal operation of a CHP plant for space heating as a peak load regulation plant. Energy vol25 pp283-298

Mathworks 1998. MATLAB 5 User's Guide

Mathworks 1998. MATLAB Optimization Toolbox, Users Guide Version 5.

Ahn, B.C, Mitchell, J.W. 2001. Optimal control development for chilled water plants using a quadratic representation. Energy and Buildings vol33 pp371-378

Dentice M. 2001. Optimal operation of a complex thermal system: a case study. Int. J. Refrigeration vol24 pp290-301

Krause A. et al. 1999. On the cost optimization of a district heating facility using a steam-injected gas turbine cycle. Energy Conversion & Management vol40; pp1617-1626

Bojic, M. et al. 2000. Mixed 0-1 sequential linear programming optimization of heat distribution in a district-heating system. Energy and Buildings vol32 pp309-317

Benonysson, 1995. A. Operational optimization in a district heating system. Energy Conversion & Management vol36(5) pp297-314

Chapter 9

General Discussion and Conclusions

9.1 RESEARCH ORIENTED (PART I)

9.1.1 Evaluation

(i) development of an integrated HAM modeling and simulation environment

The problems caused by the difference in time scales between HVAC and building response are solved by the development of a HAMBase model in SimuLink. The hybrid model consists of a continuous part with a variable time step and a discrete part with a time step of one hour. For the HVAC installation and the room response on indoor climatic variations a continuous model is used. For the external climate variations a discrete model is used. The dynamics of the building systems where small time scales play an important role are accurately simulated as shown in figures: 2.6; 2.7; 2.8; 6.3; 6.4; A2. Furthermore, the model becomes time efficient as the discrete part uses 1-hour time steps (A typical yearly based simulation takes about 2 minutes on a Pentium III, 500 MHz computer).

The simulation environment facilitates a relatively easy integration of the PDEs based models of Comsol. This is shown in Section 4.3; 4.4.2; 7.2.4 and the appendices.

(ii) verification and validation

Verification and validation (V&V) is a continuous process. The models presented in Chapter 2, the building zone model (HAMBase) and primary systems models, already show a good agreement with respectively the ASHRAE test [ASHRAE 2001] and with measurements. Preliminary results of ongoing research projects also point in that direction. In Appendix A the HAMBase model has been subjected to new V&V exercises of the IEA Annex 41 project. The final results will be published in 2008. The validation of a more advanced controller model is provided in Appendix B1.

Chapter 3 and 4 provide PDE based models. The 1D moisture case is validated with measurements. The 2D airflow models are simulated using the directly numerical solving (DNS) technique. This approach performed well for these specific examples. In general for modeling airflow, accurately 3D models are required because in principle, turbulence cannot be accurately approximated in 2D. Furthermore, due to the very limited computer recourses compared to what is needed for DNS on the scales of buildings, often turbulence models, like for example k-ε, are needed in order to get some results at all. It should be clear that the verification and validation (V&V) of CFD is still a major issue. [Oberkampf & Trucano 2002] provide useful guidelines for designing and conducting experiments for this kind of problems. The V&V of 3D HAM models of constructions is perhaps an even larger problem due to the fact that it is much more difficult to measure the required physical quantities inside capillary-porous materials than it is in air.

(iii) evaluation of the simulation environment

Limitations

(L1) Some specific solvers such as time-dependent k-ε turbulence solvers are not available in the Comsol yet. This means that, for example, time dependent 3D airflow around buildings cannot be solved (yet);

(L2) Although it is possible to construct a full 3D integrated HAM model of the indoor air and all constructions in Comsol, the simulation time would probably be far too long to be of any practical use at this moment;

(L3) A radiation modeling toolbox is just recently available in Comsol and is not included in this research.

Drawbacks

(D1) The software package MatLab itself and basic knowledge of Matlab are required to use the models;

(D2) At this moment HAMLab is a research tool. Therefore it lacks facilities for design-oriented users such as user-friendly interfaces and user guides;

(D3) Although state-of-art solvers are present, the simulation of FEM based integrated models can easily become very computation time consuming.

Benefits

(B1) It takes advantage of the facilities of the well-maintained Matlab/SimuLink and Comsol simulation environment such as the state-of-art ODE/PDE solvers, controllers library, graphical capabilities etc;

(B2) All presented HAMLab models in this thesis are public domain;

(B3) Although not explicitly shown in this thesis, compared to other HAM models, it is relatively easy to integrate new models that are based on ODEs and/or PDEs;

(B4) The simulation environment facilitates open source modeling and if desired, models can be compiled into stand-alone applications.

9.1.2 Ongoing research driven projects

Some preliminary results of current projects are summarized. All results are expected to be published in the near future. We refer to the appendix.

Part A provides recent contributions to IEA Annex 41, including: First, an analytical verification of the hygric part of the HAMBase model. Second, a complete validation of the HAMBase model, by new experimental data. Third, the results of a driving rain simulation. Fourth, a demonstration of a full dynamic 3D heat and moisture model of a small corner construction.

Part B provides preliminary results of current projects in cooperation with (MSc) students. Included are:

(Part B1) Modeling and simulation of a high tech HVAC system of a museum. The preliminary results of this project indicate that the systems modeling approach of Chapter 2 can also be applied to high tech systems including more sophisticated controllers.

(Part B2) Application of system identification. This technique seems promising in reducing the large simulation times of (linear approximated) models with large degrees of freedom (DOF) such as 3D HAM models of constructions.

(Part B3) Evaluation of full 3D airflow and constructions models. Preliminary results show that it is possible and relatively easy to model the interaction between constructions and air in Comsol. Furthermore, it seems possible to get reliable simulation results in cases for convective airflow and forced airflow. However, due to the limited computer resources used in this research (PC with 500 MB memory), the simulated volumes were small: less than 0.1 m^3 and 2 m^3 for respectively the convective and forced airflow case.

These ongoing research projects indicate that the simulation environment as a tool for research, has already successfully been utilized for both scientific and educational purposes.

9.2 DESIGN ORIENTED (PART II)

In Chapter 5 through 8 the performance based designs of HVAC systems, controllers and set point operation strategies of multiple system components are successfully evaluated using the simulation environment. Overall it may be concluded that the simulation environment (HAMLab) is quite useful for performance-based design. In more detail, the impact on the design and actual building & systems is discussed separately for each case study.

9.2.1 Evaluation

(i) evaluation of applications

Chapter 5 provides a design for the indoor climate control of a monumental theatre. The building is currently under renovation. A control class AA was originally suggested by a third party. The impact of the study was twofold: A reconsideration of control class B and a further investigation on the conservation of the monumental interior in case of control class AA by the third party. Chapter 6 presents an indoor climate controller design for the conservation of monumental paper fragments. A technically good solution is provided. However, due to esthetical considerations, the decision makers have not decided yet what to do. Chapter 7 shows a set point operation strategy design for the conservation of a monumental church organ. The suggested climate control was realized in 2001. Since that time no problems were reported. The reader should notice that six years is to short to conclude the problem is solved permanently. Chapter 8 provides an optimal operation strategy for a hospital power plant. At this moment the power plant is used for peak production of electricity.

(ii) usability in relation with integral building assessment
The application presented in Chapter 6 shows the usability in relation with integral building assessment. In this example both well defined required performance criteria and evaluation of these criteria are provided for a specific target, i.e. the conservation of paper fragments. A more systematically approach analog to [Hendriks & Hens 2000] is left over for future research.

(iii) guideline
A preliminary guideline for design-oriented users is provided in Part C of the Appendix.

9.2.2 Ongoing design driven projects
We refer to the Appendix.
Part D provides, analog to Part B, preliminary results of current projects by (MSc) students but now for design oriented projects. The reader should notice that for these type of projects only standard models, available from the website were used by the students. Because almost no modeling and simulation guidance is given, the following results can therefore be considered as almost independently achievable by MSc students.
(Part D1) The design of a hygrostatic controller. In this case, the same building is used as the one presented in Section 2.4, but indoor air heating is now hygrostatically controlled. The validation results are quite satisfactory for this controller. Several control strategy designs were evaluated.

155

(Part D2) The thermal comfort in the house of Senate. The whole building model is used for the simulation and evaluation of several design variants.

(Part D3) The design of a more efficient hypocaust heating system for a monumental Church. In this project, a rare and complex hypocaust heating system is studied. The simulation environment is used to develop a HAM model for the building and it's heating system. Again this model is used for the simulation and evaluation of several design variants.

9.3 RECOMMENDATIONS

Research oriented

It is quite clear that the development and validation of full scale 3D HAM (including CFD) models are still problematic. The author has two recommendations on this topic. First, the development of a small size (order 1 m^3) multi zone test building that can be exposed to artificial climates, including wind, rain and irradiation, is recommended. The objective of this 'toy house' is to investigate and improve the validation of full 3D HAM models that possibly can be simulated using the simulation environment presented in the thesis. Second, the development of a classification system for different types of problems related to HAM modeling and simulation. With such a classification, perhaps some types of HAM related problems are easier recognizable and manageable.

Design oriented

The Matlab Webserver Toolbox seems promising in meeting drawbacks D1 (Matlab and some knowledge of it are required) and D2 (lack of user-friendly interfaces). Already a first study resulted in web applications including user-friendly interfaces, which are accessible by a web browser (http://archbps1.campus.tue.nl/matwebserver/). A further development is recommended especially with the focus on the usability for specialized designers.

The simulation environment seems also promising for so-called early design decision support. A systematic research on this application is advised.

Literature

Ahn, B.C. & Mitchell, J.W., 2001, Optimal control development for chilled water plants using a quadratic representation. *Energy and Buildings 33* pp371-378

Adam, C. & Andre, P., 2003, Ice storage system (ISS): Simulation of a typical HVAC primary plant equipped with an ice storage, *8TH IBPSA Conference Eindhoven*, pp47-54

Andre P., Lebrun, J. & Ternoveanu, A., 1999, Bringing Simulation to appliocation; some guidelines and practical recommendations issued from IEA-BCS Annex 30, *6TH IBPSA Conference Kyoto*, 6p

Andre, P., Kelly, N., Boreux, J., Lacote, P. & Adam, C., 2003, Different approaches for the simulation of an experimental building hosting a climate chamber devoted to artificial fog production, *8TH IBPSA Conference Eindhoven*, pp47-54

Ashino, R., Nagase, M., & Vaillancourt, R., 2000, Behind and beyond the MATLAB ODE Suite, *Report CRM-2651*

Ashok, S., 2003, Optimal cool storage capacity for load management, *Energy 28*, pp115-126

ASHRAE, 2005, *ASHRAE Handbook Fundamentals*, ISBN 1931862710

ASHRAE, 2003, *ASHRAE Handbook Applications* , ISBN 1931862222

ASHRAE, 2001, Standard method of test for the evaluation of building energy analysis computer programs, *standard 140-2001*.

ASHRAE, 1993, HVAC Secondary Toolkit: A Toolkit for Secondary HVAC System Energy Calculations, *CD*, American Society of Heating, Refrigerating and Air-Conditioning Engineers, Atlanta COLORADO 80112-5776 USA

Augenbroe, G., 2002, Trends in building simulation *Building and Environment 37*, pp891 – 902

Badescu, V., 2002, Model of a solar-assisted heat-pump system for space heating integrating a thermal energy storage unit, *Energy and Buildings 34*, pp715–726

Banks, J. & Gibson, R.G., 1997, Don't Simulate When … 10 rules for determining when simulation is not appropriate. *IIE Solutions September 1997*

Bartak, M., Baeusoleil, I., Clarke, J.A., Denev, J., Drkal, F., Lain, M., Macdonald, I.A., Melikov, A., Popiolek, Z. & Stankov, 2002, Integrating CFD and building simulation, *Building and Environment 37*, pp865 – 871

Bazjanac, V., 2004, building performance simulation as part of interoperable software environments, *Building and Environment 39*, pp 879-883

Bednar, T. 2000, Beurteilung des feuchte and waermetechnischen Verhaltens von Bauteilen und Gebaeuden – Weiterentwicklung der Mess- und rechenverfahren, *PhD Thesis*, Wien University of Technology

Bednar T., Hagentoft C.E. 2005, Analytical solution for moisture buffering effect; Validation exercises for simulation tools. *Nordic Symposium on Building Physics, Reykjavik 13-15 June 2005*, pp. 625-632

Bejan A., 1993, *Heat Transfer,* - John Wiley & Sons, Inc.

Benonysson, A., 1995, Operational optimization in a district heating system. *Energy Conversion & Management 36(5)*, pp297-314

Blezer, I., 2002, Modeling, Simulation and optimization of a heat pump assisted energy roof system. *Master thesis* (in Dutch), Univ. of Tech. Eindhoven, group FAGO.

Bojic, M., Trifunovic, N. & Gustafsson, S.I., 2000, Mixed 0-1 sequential linear programming optimization of heat distribution in a district-heating system. *Energy and Buildings 32*, pp309-317

Bojic, M. & Dragicevic, S., 2002, MILP optimization of energy supply by using a boiler, a condensing turbine and a heat pump, *Energy Conversion & Managemen 43*, pp591-608

Bourdouxhe, J-P., 1998, *Reference guide for dynamic models of HVAC equipment*, ISBN 1-883413-60-5

Bradley, D. & Kummert, M., 2005, New evolutions in TRNSYS – a selection of version 16 features, *9TH IBPSA Conference Montreal*, pp107-113

Brandemuehl, M., 1993, HVAC2 Toolkit, *CD*, Joint Center for Energy Management, University of Colorado at Boulder

Brocken, H., 1998, Moisture transport in brick masonry, *PhD thesis*, Eindhoven University of Technology

Boerstra, A.C., Raue, A.K., Kurvers, S.R., Linden, A.C., van der Hogeling, J.J.N.M. & de Dear, R.J., 2005, A new Dutch adaptive comfort guideline, *Conference on the Energy performance Brussels 21-23 September 2005*, 6p

Bring, A. Sahlin, P. & Vuolle, M., 1999, Models for Building Indoor Climate and energy Simulation, *Report IEA SHC Task 22 subtask B, Version 1.02*, KTH Stockholm

Casas, W., Prölβ, K., & Schmitz, G., 2005, Modeling of desiccant assisted air conditioning systems *4^{tH} Modelica Conference. Hamburg March 7-8 2005* pp487-496

Chen, Q. & Xu, W.,1998, A zero-equation turbulence model for indoor airflows simulation, *Energy and Buildings 28*, pp137-144

Citherlet, S., Clarke, J. A. & Hand, J., 2001, Integration in building physics simulation, *Energy and Buildings 33*, pp 451-461

Clarke, J.A., 1999, Prospects for truly integrated building performance simulation, *6TH IBPSA Conference Kyoto*, 6p

Clarke, J.A., 2001, Domain integration in building simulation, *Energy and Buildings 33*, pp303-308

Clarke, J.A., 2001a. *Energy Simulation in Building Design*, 2nd Ed. Butterworth-Heinemann

Comakli, O., Bayramoglu, M. & Kaygusuz, K., 1996, A thermodynamic model of a solar assisted heat pump system with energy storage, *Solar Energy 56/6* pp485-492

COMSOL AB, 2000, FEMLAB Version 2.0 pre, Reference Manual

Crawley, D.B., Hand, J.W. Kummert, M. & Griffith B., 2005, Contrasting the capabilities of building performance simulation programs, *9TH IBPSA Conference Montreal* pp 231- 238 & *Report VERSION 1.0 July 2005*

Crawley, D.B. & 12 co-authors, 2002, EnergyPlus: New, Capable, and Linked, *International Green Building Conference, November 2002, Austin, Texas*, Washington, DC: U S Green Building Council

CSTB, 2001, SIMBAD, *CD*, Building and HVAC Toolbox

Dentice, M., 2001, Optimal operation of a complex thermal system: a case study. *Refrigeration 24* pp290-301

Djunaedy, E., 2005, External coupling between building energy simulation and computational fluid dynamics, *PhD Thesis*, Eindhoven University of Technology

Ellis, M.W. & Mathews, E.H., 2002, Needs and trends in building and HVAC system design tools, *Building and Environment 37,* pp461 – 470

Enright, W.H., 2002, The design and implementation of usable ODE software, *Numerical Algorithms 31* pp125–137

Eriksson, J.B., 1999, A method to study air distribution control, *PhD thesis*, ISSN 0284-141X

Felgner, F., Agustina, S., Cladera Bohigas, R., Merz, R., & Litz, L., 2002, Simulation of thermal building behaviour in Modelica, *2TH Modelica Conference,* pp147-154

Felsmann, C., Knabe, G. & Werdin, H., 2003, Simulation of domestic heating systems by integration of trnsys in a matlab/simulink model, *6TH Conference System Simulation in Buildings Liege,* pp79-96

Felsmann, C. & Knabe, G., 2001. Simulation and optimal control strategie of HVAC systems, *7TH IBPSA Conference Rio de Janeiro*, pp1233-1239

Fu, L., & Jiang, Y., 2000, Optimal operation of a CHP plant for space heating as a peak load regulation plant, *Energy 25*, pp283-298

Garcia-Sanz, M., 1997, A reduced model of central heating systems as a realistic scenario for analyzing control strategies, *Appl. Math modelling 21*, pp535-545

Ginestet, S., Stabat., P. & Marchio, D., 2003, Control design of open cycle desiccant cooling systems using a graphical environment tool, *6TH Conference System Simulation in Buildings Liege*, pp375-390

Gouda, M.M., Underwood, C.P. & Danaher, S., 2003, Modeling the robustness properties of HVAC plant under feedback control, *6TH Conference System Simulation in Buildings Liege*, pp511-524

Gough, M.A. 1999, A review of new techniques in building energy and environmental modeling, *BRE Report No. BREA-42, April 1999*

Hagentoft, C-E., 2002a, HAMSTAD WP2 Modelling, Version 4. *Report R-02:9. Gothenburg: Chalmers University of Technology*

Hagentoft, C-E, 2002b, Final report: Methodology of HAM-Modelling. *Report R-02:8. Gothenburg: Chalmers University of Technology*

Hanke, M., 2004, Benchmarking FEMLAB 3.0a: Laminar flows in 2D, *Report No. 2004:01, Royal Institute of Tech. Uppsala University*

Haupl, P., Jurk, K. & Petzold, H., 2003, Inside thermal insulation for historical facade, *2ND Conference on Building Physics Leuven*, pp463-469

Haupl, P. & Fechner H., 2003, Moisture behavior within the cupola of the Çhurch of Our Lady' in Dresden Germany, *2ND Conference on Building Physics Leuven,* pp785-791

Hendriks L., & Hens H., 2000, Building Envelopes in a Holistic Perspective, IEA-Annex 32, Uitgeverij ACCO, *ISBN 90-75-741-05-7*

Hendriks, L. & Linden, K. van der, 2003, Building envelopes are part of a whole: reconsidering tradional approaches, *Building and environment 38*, pp 309-318.

Hens, H., 1996, Heat, Air and Moisture Transfer in Insulated Envelope Parts, *Final report, Vol. 1, Task 1: Modelling IEA ANNEX 24*

159

Hensen, J., Djunaedy, E., Radosevic, M. & Yahiaoui, A., 2004, Building performance for better design: some issues and solutions, *21TH PLEA Conference Eindhoven*, pp1185-1190

Henze, G.P., 2003, Guidelines for improved performance of ice storage systems, *Energy and Buildings 35*, pp111-127

Hirunlabh, J., Wachirapuwadon, S., Pratinthong, N. & Khedari, J., 2001, New configurations of a roof solar collector maximizing natural ventilation, *Building and Environment 36*, pp383-391

Holm, A., Kuenzel, H.M. & Sedlbauer, K., 2003, The hygrothermal behaviour of rooms: Combining thermal building simulation and hygrothermal envelope calculation, *8TH IBPSA Conference Eindhoven*, pp499-505

Holm, A. & Kunzel, H.M., 2003, Two-dimensional transient heat and moisture simulations of rising damp with WUFI, *2ND Conference on Building Physics Leuven*, pp363-367

Hong T., Chou, S.K. & bong, T.Y., 2000, Building simulation: an overview of developments and information sources, *Building and Environment 35*, pp 347-361

Hudson, G. & C. P., Underwood, 1999, A simple building modelling procedure for MatLab/SimuLink, *6TH IBPSA Conference Kyoto*, 6p

IEA Annex 41, 2006, http://www.ecbcs.org/annexes/annex41.htm

Ihm, P., Krarti, M. & Henze, G., 2003, Integration of a thermal energy storage model within Energy plus, *8TH IBPSA Conference Eindhoven*, pp531-538

Jacobs, P. & Henderson, H., 2002, State-of-art Review of Whole building, Building Envelope and HVAC Component and systems simulation and deign tools, *Report ARTI-21CR-605-30010-30020-01*, Air-Conditioning and Refrigeration Technology Institute (ARTI)

Jaluria, Y. 1998, *Design and optimization of thermal systems*, McGraw-Hill

Jreijiry, D., Husaunndee, A., Inrad, C. & Villenave, J.G., 2003, Control of ventilation in buildings using SIMBA building and HVAC toolbox, *8TH IBPSA Conference Eindhoven*, pp591-597

Jong, J. de, Schijndel, A.W.M. van, & Pernot, C.E.E., 2000, Evaluation of a low temperature energy roof and heat pump combination, *1ST Building Physics Conference Eindhoven*, pp99-106

Judkoff, R.D. & Neymark, J.S., 1995, Building Energy Simulation Test (BESTEST) and diagnostic method. *Report NREL/TP-472-6231*, Golden, Colo. National renewable Energy Laboratory.

Karlsson, H., Sasic Kalagasisis, A. & Hagentoft C-E., 2005, Development of a modular toolbox in SimuLink for dynamic simulations of voc-concentration indoor air, *9TH IBPSA Conference Montreal*, pp493-499

Kelly, N., & Beausoleil-Morrison, I., 2003, Simulation of modern heat and power systems for buildings, *6TH Conference System Simulation in Buildings Liege*, pp307-321

Khan, M.H., 2003, Hybrid ground source heat pump system simulation using visual modeling tool for hvacsim+, *8TH IBPSA Conference Eindhoven*, pp641-648

Khoury El, Z., Riederer P., Couillland N., Simon J. & Raguin M., 2005, A multizone building model for Matlab/SimuLink environment, *9TH IBPSA Conference Montreal*, pp525-532

Klein, S.A. & Alvardo, F.L., 2002, Engineering equation solver, *CD*, F-chart software 1992-2002

Knabe, G. & Le, H-T, 2001, Building simulation by application of a HVAC system considering the thermal and moisture behaviors of the perimeter 7TH *IBPSA Conference Rio de Janeiro*, pp965-972

Knabe, G. & Le, H-T, 2003, Influence of controller types in HVAC split systems on zone comfort and energy consumption, 8TH *IBPSA Conference Eindhoven*, pp649-656

Künzel, M., 1995, Verfahren zur einund zweidimensionalen Berechnung des gekoppelten Wärmeund Feuchtetransports in Bauteilen mit einfachen Kennwerten. *PhD Thesis*, IRB Verlag, Stuttgart

Kunzel, H.M., 1996, *IEA Annex 24 HAMTIE, Final Report - Task 1*

Kolsaker, K., 1994, NEUTRAN – A translator of models from NMF into IDA and SPARK. *BEPAC Conference. "BEP'94". York*, pp1120–1128

Kowalski, S.J. & Rybicki, A., 1999, Computer Simulation of Drying Optimal Control, *transport in Porous Media 34*, pp227-238

Kowalski, S.J., 2002, Modeling of fracture phenomena in dried materials, *Chemical Engineering Journal 86*, pp145-151

Krause, A., 1999, On the cost optimization of a district heating facility using a steam-injected gas turbine cycle, *Energy Conversion & Management 40*, pp1617-1626

Lebrun, J., Bourdouxhe, J.-P. & Grodent, M., 1994 -A toolkit for primary HVAC system energy calculation, *CD*, prepared for the American Society of Heating, Refrigerating and Air Conditioning Engineers - TC 4.7 Energy Calculation Laboratoire de Thermodynamique, Université de Liège, Belgium

Liao, Z. & Dexter, A.L., 2004, A simplified physical model for estmating the average air temperature in multi-zone heating systems, *Building and environment 39*, pp 1013-1022.

Lillicrap D.C. & Davidson P.J., 2005, Building energy standards – tool for certification (bestcert) – pilot methodologies investigated. *Conference on the Energy performance Brussels 21-23 September 2005*, 6p

Loomans, M.G.L.C. & Schijndel, A.W.M. van, 2002, Simulation and measurement of the stationary and transient characteristics of the hot sphere anemometer, *Building and Environment 37*, pp153-163

Mahdavi, A., 2004, Reflections on computational models, *Building and Environment 39*, pp913-925

Malkawi, A.M. & Augenbroe, G., 2004, *Advanced Building Simulation*. Spon Press, ISBN 0-415-32122-0

Massie, D.D., 2002, Optimization of a building's cooling plant for operating cost and energy use, *Themal Sciences 41*, pp1121-1129

Mathworks, The, Inc. 1998, *MatLab Manual*, Version 5.3.

Mathworks, The, Inc. 1999, *Simulink Reference Manual*, Version 3.

Mathworks, The, Inc. 1999, MATLAB Optimization Toolbox, *Users Guide* Version 5

Mechaqrane, A. & Zouak, M., 2004, A comparison of linear and neutrl network ARX models applied to a prediction of the indoor temperautre of a building. *Neural Computing & Applicication 13*, pp32-37.

Mendes, N., Oliveira, R.C.L.F. de & Santos, G.H. dos, 2001, DOMUS 1.0: A brazilian pc program for building simulation, 7TH *IBPSA Conference Rio de Janeiro*, pp83-89

Mendes, N., Winkelmann, F.C., Lamberts, R. & Phillipi, P.C., 2003, Moisture effects on conduction loads, *Energy and Buildings 35*, pp257-271

Mendes, N., Oliveira, R.C.L.F., Araujo, H.X. & Coelho, L.S., 2003a, A matlab-based simulation tool for building thermal performance analysis, 8TH *IBPSA Conference Eindhoven*, pp855-862

Mendes, N., Oliveira, R.C.L.F. de & Santos, G.H. dos, 2003b, DOMUS 2.0: A whole-building hygrothermal simulation program, 8TH *IBPSA Conference Eindhoven*, pp863-870

Moon, H.J. & Augenbroe, G., 2003, Evaluation of hygrothermal models for mold growth avoidance prediction, 8TH *IBPSA Conference Eindhoven*, pp895-902

Mora, L., Mendonca, K.C., Wirtz, E. & Inard, C., 2003, Simspark: an object-oriented environment to predict coupled heat and mass transfer in buildings, 8TH *IBPSA Conference Eindhoven*, pp903-910

Murakami, S., Kato, S. & Kim, T., 2001, Indoor climate design based on CFD Coupled simulation of convection, radiation, and HVAC control for attaining a given PMV value, *Building and Environment 36*, pp701–709

Musy, M., Wurtz, E., Winkelmann, F. & Allard, F., 2001, Generation of a zonal model to simulate natural convection in a room with radiative/convective heater, *Building and Environment 36*, pp589-596

Neilen, D., Schellen, H.L. & Aarle, M.A.P. van, 2003, Characterizing and comparing monumental churches and their heating performance, 2ND *Conference on Building Physics Leuven*, pp793-801

Neuhaus, E. & Schellen, H.L., 2006, Onderzoek hygrostatisch geregeld stoken GEOhuis (in Dutch), *BPS-report 06.02.K*, Eindhoven University of Technology.

Neymark, J., Judkoff, R., Knabe, G., Le, H.-T., Durig, M., Glass, A. & Zweifel, G., , 2002, Applying the building energy simulation test (BESTEST) diagnostic method to verification of space conditioning equipment models used in whole-building energy simulation programs, *Energy and Buildings 34*, pp917–931

Ngendakumana, P. 2003, Solving rating and sizing problems of cross-flow heat exchanger with both fluids unmixed by means of engineering equation solver, 6TH *Conference System Simulation in Buildings Liege*, pp185-196

Novak, P.R., Mendes, N. & N. Oliveira G.H.C, 2005, Simulation of HVAC plants in 2 brazilian cities using Matlab/SimuLink, 9TH *IBPSA Conference Montreal*, pp859-866

Nytsch-Geusen, C., Klempin, C., Nunez, J. & Radler, J., 2003, Integration of CAAD, thermal building simulation and CFD by using the ifc data exchange format, 8TH *IBPSA Conference Eindhoven*, pp967-973

Nytsch-Geusen, C.N., Nouidui, T. Holm, A. & Haupt W., 2005, A hygrothermal building model based on the object-orented modeling language modelica. 9TH *IBPSA Conference Montreal*, pp867-873.

Oberkampf, W.L., DeLand, S.M., Rutherford, B.M., Diegert, K.V. & Alvin K.F., 2002, Error and uncertainty in modeling and simulation. *Reliability Engineering and System Safety 75*, pp333-357

Oberkampf, W.L. & Trucano, T.G., 2002, Verification and validation in computational fluid dynamics, *Progress in Areospace Sciences 38*, pp209-272

Olsson, H., 2005, External Interface to Modelica in Dymola, *4^{tH} Modelica Conference. Hamburg,* pp603-611

Pohl, S.E. & Ungethum, J., 2005, A simulation management environement for Dymola. *4^{tH} Modelica Conference. Hamburg,* pp. 173-176.

Prazers, L. & Clarke, J.A., 2003, Communicating building simulation outputs to users, *8TH IBPSA Conference Eindhoven,* pp1053-1060

Randriamiarinjatovo, D., 2003, Building energy simulation with EASE, *8TH IBPSA Conference Eindhoven,* pp1085-1092

Ren, Z. & Stewart, J., 2003, Simulating air flow and temperature distribution inside buildings using a modified version of COMIS with sub-zonal divisions, *Energy and Buildings, 35,* pp303-308

Riederer, P., Marchio, D., Visier, J.C., Husaunndee, A. & Lahrech, R., 2002, Room thermal modelling adapted to the test of HVAC control systems, *Building and Environment 37,* pp777 – 790

Riederer, P., 2003, From sizing and hydraulic balancing to control using the SIMBAD toolbox, *8TH IBPSA Conference Eindhoven,* pp1101-1108

Riederer, P. & Marchio, P. 2003a, Sensitivity analysis of a new room model integrating phenomena of air temperature distribution in controller tests, *6TH Conference System Simulation in Buildings,* pp39-54

Rieder P., 2005, MatLab/SimuLink for building and HVAC simulation – state of art. *9TH IBPSA Conference Montreal,* pp1019-1026

Rode, C., Salonvaara, M., Ojanen, T., Simonson, C. & Grau, K., Integrated hygrothermal analysis of ecological buildings, *2ND Conference on Building Physics Leuven,* pp859-868

Sahlin, P., 1996, Modelling and Simulation Methods for Modular Continuous Systems in Buildings. *PhD Thesis.* KTH, Stockholm

Sahlin, P., 1996, NMF Handbook. An Introduction to the Neutral Model Format, NMF Version 3.02. *ASHRAE Report 839,* Dept. of Building Sciences, KTH, Stockholm

Sahlin, P., 2003, Will equation based building simulation make it? Experiences from the introduction of ida indoor climate and energy, *8TH IBPSA Conference Eindhoven,* pp1147-1154

Sahlin, P., Eriksson, L., Grozman, P., Johnsson, H., Shapovalov, A. & Vuolle, M., 2004, Whole-building simulation with symbolic DAE equations and general purpose solvers, *Building and Environment 39,* pp949-958

Saldamli, L., Bachmann, B., Fritzson, P. & Wiesmann, H. 2005. A framework for describing and solving PDE Models in Modelica, *4^{tH} Modelica Conference Hamburg,* pp113-122.

Salsbury, T. & Diamond, R., 2000, Performance validation and energy analysis of HVAC systems using simulation, *Energy and Buildings 32,* pp5-17

Sasic Kalagasidis, A., 2004, HAM-Tools, An Integrated Simulation Tool for Heat, Air and Moisture Transfer Analyses in Building Physics, *PhD Thesis,* ISSN 1400-2728

Sasic Kalagasidis, A. & Hagentoft, C-E., 2003, The influence of air transport in and around the building envelope on energy efficiency of the building, *2ND Conference on Building Physics Leuven,* pp729-735

163

Schellen, H.L., 2002, Heating Monumental Churches, Indoor Climate and Preservation of Cultural heritage; *PhD Thesis*, Eindhoven University of Technology.

Schellen, H.L., Schijndel, A.W.M. van, Neilen, D. & Aarle, M.A.P. van, 2003, Damage to a monumental organ due to wood deformation caused by church heating, 2ND *Conference on Building Physics Leuven*, pp803-811

Schijndel, A.W.M. van & Schellen H.L. 2005a. Application of an integrated indoor climate & HVAC model for the indoor climate performance of a museum, 7TH *Symposium on Building Physics in the Nordic Countries Reykjavik*, pp1118-1125

Schijndel, A.W.M. van & Hensen, J.L.M., 2005b, Integrated heat, air and moisture modeling toolkit in Matlab. 9TH *IBPSA Conference Montreal*, pp1107-1114

Schijndel, A.W.M. van, 2005c, Implementation of FemLab in S-Functions, 1ST *FemLab Conference Frankfurt*, pp324-329.

Schijndel, A.W.M. van, 2005d, Application of an integrated heat, air and moisture modeling toolkit in MatLab for design, *IBPSA-NVL Conference Delft*, 8p

Schijndel, A.W.M. van, 2005e, Indoor climate design for a monumental building with periodic high indoor moisture loads, 26TH *AIVC Conference Brussels*, pp301-313

Schijndel, A.W.M. van, 2003a, Modeling and solving building physics problems with FemLab, *Building and Environment 38*, pp319-327

Schijndel, A.W.M. van, Schellen, H.L., Neilen, D. & Aarle, M.A.P. van, 2003b, Optimal setpoint operation of the climate control of a monumental church, 2ND *Conference on Building Physics Leuven,* pp777-784

Schijndel, A.W.M. van & Wit, M.H. de, 2003c, Advanced simulation of building systems and control with SimuLink, 8TH *IBPSA Conference Eindhoven*, pp1185-1192

Schijndel, A.W.M. van, 2003d, Integrated building physics simulation with FemLab/SimuLink/Matlab, 8TH *IBPSA Conference Eindhoven*, pp1177-1184

Schijndel, A.W.M. van, 2003e, Advanced HVAC modeling with FemLab/SimuLink/Matlab, *Building Serv. Eng. Res. Technol. 24:4*, pp289-300

Schijndel, A.W.M. van, 2002a, Advanced HVAC modeling with FemLab/SimuLink/MatLab, 6TH *Conference System Simulation in Buildings Liege*, pp 243-260

Schijndel, A.W.M. van & Wit, M.H. de, 2002b, Advanced system and control simulation of an energy roof system. *11th Symposium for building physics Dresden*, pp180-189

Schijndel, A.W.M. van, 2002c, Optimal operation of a hospital power plant, *Energy and Buildings 34*, pp1055-1065.

Schijndel, A.W.M. van, 2002d, Solving building physics problems based on PDEs with FemLab, 7TH *Symposium on Building Physics in the Nordic Countries Trondheim*, pp9-16

Schijndel, A.W.M. van & Wit, M.H. de, 1999a, A building physics toolbox in MatLab, 7TH *Symposium on Building Physics in the Nordic Countries Goteborg*, pp81-88

Schijndel, A.W.M. van, 1999b, Setpoint Optimization of a power plant, 2ND *Benelux MatLab Usersconference Brussels*, Chapter 3

Schijndel, A.W.M. van, 1998, Optimization of the operation of the heating, power and cooling installation of the Academic Hospital Groningen, *report no. NR-2029*, Eindhoven University of Technology

Schijndel, A.W.M. van, 1997, Building physics applications in Matlab, I^{ST} *Benelux MatLab Usersconference Amsterdam*, Chapter 11

Schwab, M. & Simonson, C., 2004, Review of building energy simulation tools that include moisture storage in building materials and HVAC systems, Draft Report IEA Annex 41, Zurich

Sinha, S.L., Arora, R.C. & Subhransu, R., 2000, Numerical simulation of two-dimensional room air flow with and without buoyancy, *Energy and Buildings 32,* pp121-129

Sowell, E.F.,& Moshier, M.A., 2001, Efficient solution strategies for building energy system simulation, *Energy and Buildings 33*, pp309-317

Sowell, E.F. & Moshier, M.A.,2003 Application of the spark kernel, 8 TH IPBSA Conference Eindhoven, pp1235-1242

Stappers, M.H.L. 2000. De Waalse Kerk in Delft; onderzoek naar het behoud van een monumentaal orgel (in Dutch), *MSc thesis*, FAGO 00.03.W, Eindhoven University of Technology.

Stec, W. & Paaasen, D., van, 2003, Defining the performance of the double skin facade with the use of the simulation model, 8 TH *IPBSA Conference Eindhoven*, pp1243-1249

Steskens, P.W.M.H., 2006, The advanced modelling and simulation of complex heat air and moisture interactions in buildings. *MSc thesis*, BPS, Eindhoven University of Technology.

Straube, J.F & Schumacher, C.J., 2003, Hygrothermal enclosure models: comparison with field data, 2ND *Conference on Building Physics Leuven,* pp319-325

Teodosiu C., Hohota, R., Rusaouen, G. & Wloszyn, M., 2003, Numerical prediction of indoor air humidity and its effect on indoor environment, *Building and Environment 38*, pp655 – 664

Timmermans, W.J., 2006, Modelvorming van een geavanceerd HVAC systeem van een museum depot (in Dutch), *MSc thesis*, Eindhoven University of Technology.

Underwood, C.P., 1999, *HVAC Control Systems*, ISBN 0 419 20980 8

Verdier, O., 2004, Benchmark of FemLab, Fluent and Ansys, *Preprints in mathematical sciences 2004:6*, Lund University

Virk G.S., Azzi, D., Azad, A.K.M. & Loveday D.L., 1998, Recursive models for multi-roomed bms applications. *UKACC Conference on Control 1-4 september 1998*, pp1682-1687

Weber, A., Koschenz, M., Dorer, V., Hiller, M. & Holst, S., 2003, Trnflow, a new tool for the modeling of heat, air and pollutant transport in buildings within trnsys, 8 TH *IPBSA Conference Eindhoven*, pp1363-1367

Weitzmann, P. Sasic Kalagasidis, A., Rammer Nielsen, T. Peuhkuri, R. & Hagentoft C-E., 2003, Presentation of the international building physics toolbox for SimuLink*, 8 TH IPBSA Conference Eindhoven*, pp1369-1376

Welfonder, T., Hiller, M., Holst, S. & Knirsch A., 2003, Improvement on the capabilities of trnsys15, 8 TH *IPBSA Conference Eindhoven*, pp1377-1383

Wiechert,W., 2003, The role of modeling in computational science education, *Future Generation Computer Systems 19*, pp1363–1374

Wijffelaars, J.L. & Zundert, K. van, Behouden of verouderen – Onderzoek naar het binnenklimaat en de behangfragmenten in de kamer van Anne Frank (in Dutch), *MSc thesis*, FAGO 04.032.W, Eindhoven University of Technology.

Wit M.H. de & Driessen, H.H., 1988, ELAN A Computer Model for Building Energy Design. *Building and Environment 23*, pp.285-289

Wit, M.H. de, 2006, HAMBase, Heat, Air and Moisture Model for Building and Systems Evaluation, Bouwstenen 100, ISBN 90-6814-601-7 Eindhoven University of Technology

Yahiaoui, A., Hensen, J. Soethout, L. & Paassen, D, 2005, Interfacing building performance simulation with control modeling using internet sockets, *9TH IBPSA Conference Montreal 2005*, pp1377-1384

Yu B, & Paassen, A.H.C. van, 2003, Simulink and bond graph modeling of an air-conditioned room, Simulation *Modelling Practice and Theory Volume 12*, pp61-76

Yu, B., 2003a Level-Oriented Diagnosis for indoor Climate Installations, *PhD thesis*, ISBN 90-9017472-9

Yu, B., Jong, W. de & Paassen, A.H.C. van, 2003b, General modeling of fin-tube heat exchanger of ahu, *6TH Conference System Simulation in Buildings Liege,* pp211-220

Zhai, Z., Chen, Q., Haves, P. & Klems, H., 2002, On approaches to couple energy simulation and computational fluid dynamics programs, *Building and Environment 37*, pp 857 – 864

Zhai Z. & Chen, Q., 2003, Solution of iterative coupling between energy simulation and CFD programs, *Energy and Buildings 35*, pp 493-505

Zhai Z. & Chen, Q., 2005, Performance of coupled building energy and CFD, *Energy and Buildings 37*, pp 333-344

Zhao, H. Holst, J. & Arvaston, L., 1998, Optimal operation of coproduction with storage, *Energy 23*, pp859-866

Zhao, B., Li, X. & Yan, Q., 2003, A simplified system for indoor airflow simulation, *Building and Environment 38*, pp543-552

Index of models

The next index provides an overview of the models used in the thesis.

Abbreviations:

Val: validation level: (M)easurements; (A)nalytical; (C)omparison with other models; (D)emonstration

Type: type of model: (H)eat ; A(ir) ; (M)oisture

Dim: model dimension: (0123D)

time: timestep :   (cont)inuous; (hr) step; (stat)ic

Acronym	Title	Val	Type	Dim.	time	Section
Church1	Small church indoor climate	M	HAM	0D	cont	7.2.1
Church2	Large church indoor climate	M	HAM	0D	hr	D3
Control1	On/off RH Control	D	HAM	0D	cont.	2.3
Control2	P controller & HVAC	D	HAM	0D	cont.	6.4.1
Control3	P controller & heater	D	HAM	0D	cont	7.2.3
Control4	Temperature change rate limiter	D	HAM	0D	cont.	7.4.1
Control5	RH change rate limiter	D	HAM	0D	cont.	7.4.2
Control6	Hygrostatic control	M	HAM	0D	cont.	D1
HAMBase	Whole building performance	MC	HAM	0D	hr	2.2
HAMBaseS	Whole building performance	MC	HAM	0D	cont.	2.3
HeatMois1	Heat & moisture transport in brick	D	HM	3D	cont.	3.4
HeatMois2	Heat & moisture transport in corner	D	HM	3D	cont.	A3
HVAC1	Primary HVAC system	D	H	0D	cont	2.5
HVAC2	Indoor climate & HVAC system	M	HAM	0D	cont	6.3.2
HVAC3	high tech system & controllers	DM	HAM	0D	cont	B1
Hygric	Hygric indoor climate	A	MA	1D	cont.	A
Moist1	Moisture transport in brick	M	M	1D	cont.	3.3.1

Appendix A.

IEA Annex 41 preliminary results
(Final reports expected in 2008)

Verification study

The hygric part of HAMBase has been subjected to new benchmark cases developed by the IEA Annex 41. The geometry is identical to the one in the [ASHRAE 2001]. Further test conditions are as follows: (1) there is only one 1D construction with linear properties; (2) the exposure is completely isothermal; (3) the outdoor relative humidity (RH) is 30%; (4) there are no windows; (5) the internal gains equal 500g/hour; (6) the ventilation rate is 0.5 ach. This study presents three solutions of the benchmark: (1) An (semi)exact solution, obtained by the integration of a Comsol model of the construction into a first order model of the room modeled in SimuLink as presented in the next figure:

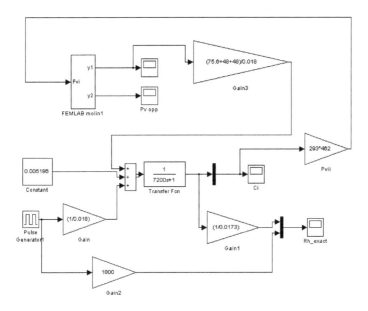

Figure A1. The SimuLink model of the (semi) exact solution

(2) Solutions obtained by HAMBase (hourly values) and (3) by HAMBase SimuLink (continuous value). Figure A2 and A3 show the results of two benchmarks representing moisture buffering by air alone and by combined air and the construction.

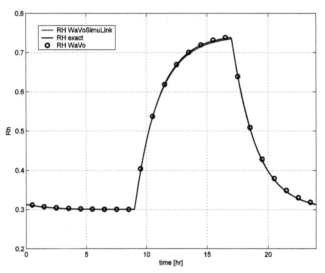

Figure A2. Simulated RH of the indoor climate with only moisture buffering by air

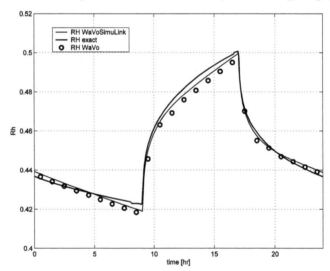

Figure A3. Simulated RH of the indoor climate with moisture buffering by air and the construction

The (semi) exact solutions are identical to the analytical solutions provide by [Bednar & Hagentoft 2005]. Furthermore, our semi exact solution was obtained before the publication of the analytical solution. The solutions obtained by the HAMBase and HAMBase SimuLink models are satisfactory.

Validation study

The intention of this study was to simulate two real test rooms, which are located at the outdoor testing site of the Fraunhofer-Institute of building physics in Holzkirchen. During the winter of 2005-2006, tests were carried out with the aim to compare the measurements with the models, developed within the Annex 41. In the reference room a standard common used gypsum plaster with a latex paint is used. The walls and the ceiling of the test room are fully coated with aluminium foil. The next results are presented: the measured and simulated RH in the reference room (A4), test room (A5) and heating power to the reference room (A6) and test room (A7).

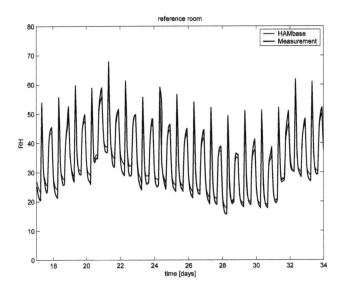

Figure A4. The measured and simulated RH in the reference room

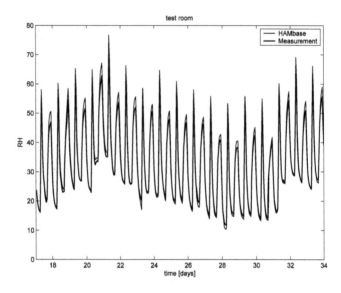

Figure A5. The measured and simulated RH in the test room

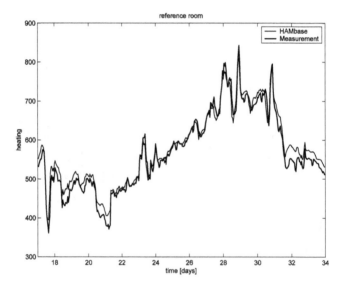

Figure A6. The measured and simulated heating power in the reference room

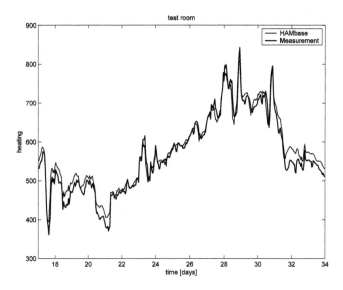

Figure A7. The measured and simulated heating power in the reference room

Again, solutions obtained by HAMBase are satisfactory.

Wind and driving rain study

Another benchmark case is related with boundary conditions. For several building geometries, it is requested to: (1) Simulate (3D) wind velocity profiles around buildings; (2) simulate raindrop trajectories and (3) simulated wind-driving-rain coefficients on the building facades.

A 3D solution of (1) could not be obtained using Comsol due to the limited capabilities of the current solvers. However, a 2D solution could be obtained using the standard k-ε turbulence model of Comsol. This presents a symmetric plane at the centre of the (3D) building. Although, it is notified that in principle 3D turbulence cannot be modelled correctly in 2D, it is interesting to investigate the accuracy of 2D turbulent models compared to 3D. Figure A8 shows the simulated 2D wind profile around the lower benchmark building.

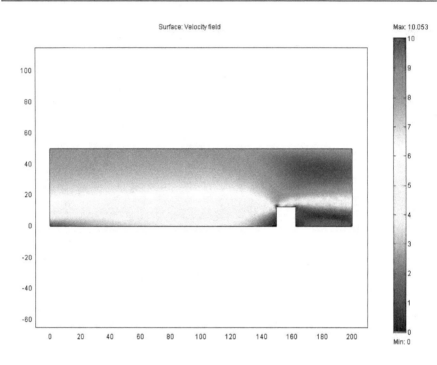

Figure A8. The wind velocity profile: absolute value of the velocity [m/s] for each coordinate x, y [m].

The solution of (2) is obtained from the equation of motion of a raindrop, moving in a wind-flow field characterized by a velocity vector v:

$$g + f(Cd, \mathrm{Re}) \cdot (v - \frac{dr}{dt}) = \frac{d^2 r}{dt^2} \tag{A1}$$

where g is gravity, f is a function dependent on Re (Reynolds number) and Cd (drag coefficient), v is wind velocity, t is time and r is the position of raindrop. Equation A1 is solved with the ODE solver of Matlab providing raindrop trajectories. The result for the low building benchmark case is shown in figure A9.

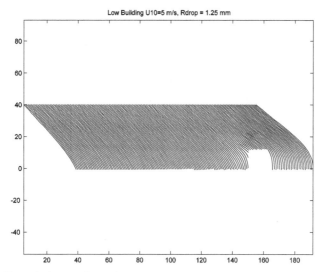

Figure A9. The raindrop trajectories

The solution of (3) is obtained from the locations of raindrops hitting the building façade. Figure A10 presents a comparison of our results (2D CFD) with the results by another 3D model (Fluent), 2D potential flow and calculation methods (PrEN and DRF-RAF)

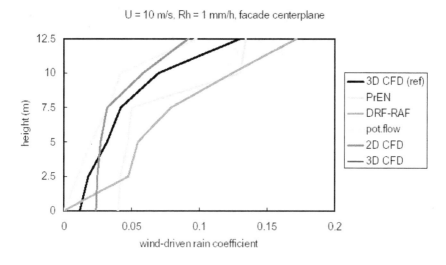

Figure A10. Wind-driven rain coefficient at the centre of the building

Full 3D HAM modeling study

This Section demonstrates the integration of a 3D HAM model of a roof/wall construction into a whole building model. The geometry of the building is identical to one in [ASHRAE 2001]. HAMBase SimuLink is used as whole building model. Figure A11 shows the input/output structure of the integrated model.

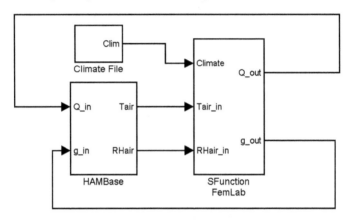

Figure A11. Input/output structure of the SimuLink model

The output (air temperature and RH) of this model and climate data are used as input for the Comsol model of a 3D roof/wall construction. The output of the Comsol model (heat and moisture flow) is connected to the input of the building model. The Comsol model is similar to the model presented in Section 3.4. Figure A12 and A13 demonstrate the 3D temperature and vapor distributions at a specific time.

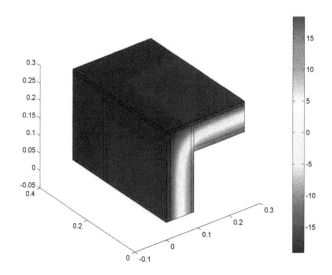

Figure A12. The temperature distribution: temperature [^{o}C] for each coordinate x, y, z [m].

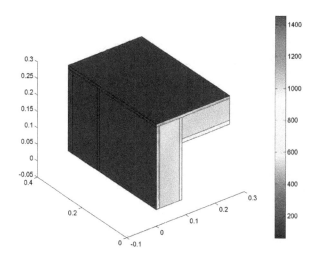

Figure A13. The vapor pressure distribution: vapor pressure [Pa] for each coordinate x, y, z [m].

177

The integrated model is capable of simulating the indoor climate and construction details within reasonable simulation time: 1 year simulation, takes about 12 hours of computation time (Pentium 4, 500MB)

Acknowledgement
The contribution of Martin de Wit to the verification and validation of HAMBase is greatly acknowledged.

Appendix B. Ongoing research projects

B1. The modeling of a high tech HVAC system of a museum

From paper by A.W.M. van Schijndel & W.J. Timmermans, accepted for publication at 7TH SSB Conference Liege 2006 December

The study concerns the HVAC system of the National Naval Depot, which should have a very high reliability. However, during the year a seemingly harmless HVAC fault almost caused a serious problem for the preservation of the artifacts. As a result of this, the next research questions are investigated in this project. What is the performance of this high tech installation in case of a major failure? Is it possible to improve the climate control in such a case? The methodology of research was: First, we implemented heat, air & moisture (HAM) models of the building and installation components in SimuLink. Second, we validated the models by measurements. Third, we evaluated the current and new designs by simulation. In [Timmermans 2006], the following results are presented in more detail: (1) Evaluation of the current HVAC system components and indoor climate of the museum; (2) Evaluation of validation results; (3) Evaluation of the simulated performance of the current design in case of failure; (4) The performance of improved designs in case of a failure. It is concluded that the current design performs well if in case of a fault, the air supply to the depots is switched off automatically. The construction of the depots contains sufficient thermal inertia to maintain a stable indoor climate for a longer period in which the fault can be repaired. A further improvement of the design could be to control the climate surrounding the depots instead of controlling the indoor climate in the depots itself. In this case, even if the system would not detect a fault and thus supplies uncontrolled air at the surroundings of the depot, the indoor climate in the depot would remain stable.

Exemplary results

Figure B1 shows the HVAC system including the cooling coil. Figure B2 presents the measured air temperature before the cooling coil and the measured and simulated air temperature after the cooling coil.

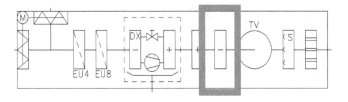

Figure B1 HVAC system and cooling coil

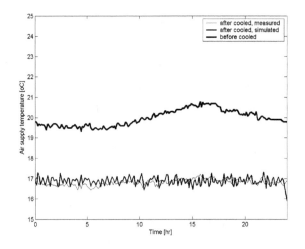

Figure B2 Air temperature before and after the cooling coil.

Context to the thesis

The preliminary results of this project indicate that systems modeling approach of Chapter 2 can also be applied to high tech systems including more sophisticated controllers.

Acknowledgement

Walter Timmermans is acknowledged for his contribution

180

B2. System Identification

From paper by P.W.M.H. Steskens & A.W.M. van Schijndel accepted for publication at
7TH SSB Conference Liege 2006 December

The study focuses on the applicability of system identification to identify building and system dynamics for climate control design. The main problem regarding the simulation of the dynamic response of a building using building simulation software is that simulation results often lack information regarding fast dynamic behavior (in the order of seconds) of the building, since most software uses a discrete time step, usually fixed to one hour. However, information about fast dynamic behavior and transient responses, which often lie within this time step, may be essential for application of an appropriate control strategy such as on/off control. The paper has two objectives. The first objective is to study the applicability of system identification to identify fast building dynamics based on discrete time data (one hour). A second objective is to research the applicability of the system for climate control design, focusing on fast dynamics (in the order of seconds). First, a toolbox is selected for model identification of building dynamics and the quality of the results produced by this toolbox is verified. Second, the applicability of system identification for internal temperature control is researched based on a simulation with free-floating indoor air temperature and a simulation with on/off-controlled indoor air temperature. Third, model identification based on similar simulations of HAM processes has been evaluated. Fourth, external data, retrieved from a building performance simulation, has been used to identify building dynamics. The study illustrates that the used system identification tool seems to be limited in accurate identification of continuous linear systems out of discrete (hourly) data due to the lack of fast dynamics in the hourly data.

Identification of discrete linear systems out of discrete (hourly) data seems to be possible if these data contain enough 'impacts' in the time series, which can be indicated by the so-called Crest Factor.

Exemplary results

The next figures show the linear system approach (B3) and the result (B4).

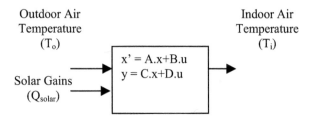

Figure B3 Continuous time state-space model of the building dynamics.

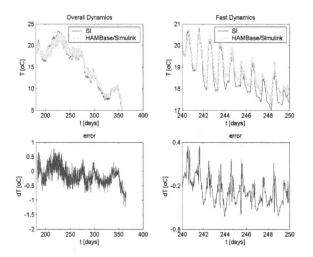

Figure B4 Comparison between the indoor air temperatures predicted by the SI model and the HAMBase/Simulink model.

Context to the thesis

This technique is promising in solving drawback D3 and limitation L2.

Acknowledgement

Paul Steskens is acknowledged for his contribution.

B3. Full 3D HAM Modeling

Abstract of corresponding MSc thesis by P.W.M.H. Steskens, 2006 June

The purpose of this research project is to develop a computational model for multi dimensional transient Heat, Air and Moisture (HAM) flows in buildings. This project intends to produce a tool that enables the analysis of conditions leading to degradation of building components. It is the intention that with a combination of a transient multidimensional Heat, Air and Moisture (HAM) building model, models describing deterioration in building materials, and systematic collection of empirical knowledge, it may be possible to predict better the degradation processes in building materials. Two problems that are related to the development of a multidimensional transient HAM building model are studied. The first issue considers the research of the local temperature and humidity levels and variations in a room. The perspectives and possibilities of modeling a room with surrounding construction are studied. First of all, the airflow in the room as well as the temperature distribution in the building construction and materials is modeled. The modeling results in a transient multidimensional model of a room, which describes the thermal conditions (Heat and Air flow) in the room. Second, the model is verified and validated using experimental data. Third, the model is extended by adding moisture flow to the model. Finally, the obtained transient multidimensional HAM model of the room is verified and validated using experimental data obtained from literature.

The second problem issue considers the research to the detailed thermodynamic behavior of a floor heating. First of all, models, describing the thermodynamic behavior of a floor heating, documented in literature and scientific articles are researched. Second, starting from a scientific article that is representative for the state-of-the-art in modeling floor heating thermal behavior, the results presented in the article have been reproduced. Third, the model is extended and improved for the general application in building services engineering and control design. The resulting model is expected to be a transient multidimensional detailed model of the thermal behavior of a floor heating system in a building.

Exemplary results

The next figures show the results of a full 3D combined room airflow & wall model (B5) and combined duct airflow & floor model (B6).

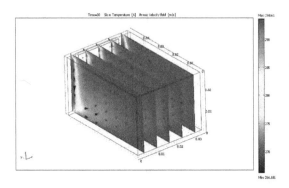

Figure B5. The temperature distribution of the air in a small room

Figure B6. Temperature distribution of a concrete floor heated with an air duct

Context to the thesis

Preliminary results show that it is possible and relatively easy to model the interaction between constructions and air in FemLab

Acknowledgement

Paul Steskens is acknowledged for his contribution

Appendix C

Preliminary Guideline

This appendix provides a step-by-step guideline. Each step requires more knowledge of the building and systems details as well as the modeling environment. After each step, the simulation results should be evaluated. If the results are not satisfactory due to oversimplification of the model(s) go to the next step.

Step 0: Selection of a similar building.

The first step neither requires Matlab nor specific knowledge of modeling parameters. At the website of HAMLab there is a web base application containing already simulated results of several building types [http://sts.bwk.tue.nl/hamcases/]. For each building type graphical output is presented including time series and statistical information of the temperature, relative humidity (RH), thermal comfort, heating, cooling and (de)humidification in each zone for a specific period (year, month week). A building should be selected that is most similar to the design.

Step 1: Editing/simulating the HAMBase model

This step requires software package Matlab R13 and basic knowledge of HAM parameters of the building. After downloading HAMBase (public domain), the selected building of step 0 can easily be simulated by typing the name of building type at the Matlab prompt. The same results as presented at the website occur. The input file is a text file that can be edited to match the design. The model parameters are explained in the same file. The HAMBase model facilitates default systems and controllers. If it is necessary to model more details of the systems and controllers go to step 2.

Step 2: Including detailed systems and controllers

This step also requires: SimuLink, included in MatLab R13, some experience with this tool and basic knowledge of systems and controllers. It is very easy to export the HAMBase model to a SimuLink block. The input/output structure of this block is as follows: for each zone the input consists of a heat and moisture source and the output consists of the air temperature, comfort temperature and RH. The exported SimuLink model simulates a free-floating situation of the building, where all systems and controller settings of the original MatLab model are ignored. Detailed system and controller models can be added in SimuLink in order to simulate a controlled situation of the building. Note that also models mentioned in Section 2.4 can be used in this stage. If it is necessary to model more details of the constructions or airflow go to step 3.

Step 3: Including detailed HAM/Airflow models

This step requires: Beside the Matlab R13 software, a separate toolbox Comsol (R3), experience with this tool and specific knowledge of PDE modeling. To integrate HAM/Airflow models into SimuLink, the next sub-steps are required: (a) Select a Comsol model that is most suitable for the problem. Inspiring models can be found at the Comsol and HAMLab websites; (b) adapt the model, start simulations and evaluate the simulation results. If satisfactory, (c) export the Comsol model to SimuLink using the standard export facility or using S-Functions as shown in Section 4.3

Step 4: Evaluation

If the final model is ready, the simulation results should be evaluated and if necessary the modeling parameters and/or the models themselves should be modified.

Appendix D. Ongoing design projects

D1. Conservational heating to control relative humidity and create museum indoor conditions in a monumental building

Abstract by H.L. Schellen & E. Neuhaus submitted to Buildings X International Conference Clearwater Beach, Florida, 2007 December

For the conservation of an important museum collection a controlled indoor climate is necessary. One of the most important factors is controlling relative humidity. These museum collections often are part of the interior of a monumental building. Originally these buildings did not have any other heating system than open fire or some kind of local heating system. Sometimes a central heating system was installed afterwards. In most cases the possibilities to fully control relative humidity, e.g. by installing a full air-conditioned system are limited due to the dimensions of the ductwork and the necessary demolishing work. Furthermore local humidifying devices may lead to dramatic indoor air conditions with condensation effects on the cold indoor surfaces of the exterior walls, single glazing and roofs. One way to overcome this problem is to make use of a so-called 'conservational heating'. A hygrostatic device to limit relative humidities controls the heating system. High relative humidity is prevented by starting heating and reaching low relative humidity will stop heating. The use of this control system is limited. In summer it may be necessary to start heating and during wintertime it may be necessary to limit heating. A comparison of the indoor climate for a usual thermostatic heating and a hygrostatic heating was gained, both by measurements in laboratory and on site. The results were compared with simulation results. A control strategy was part of the work.

Exemplary results

Figure D1 shows the results of a humidistatically controlled room of the GEO test site (see also Section 2.4 – 2.6).

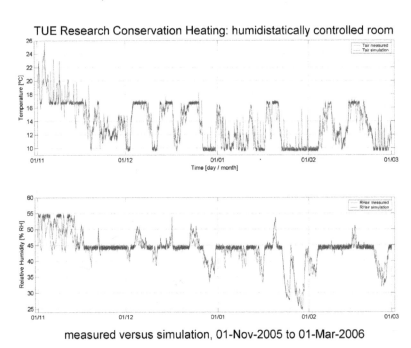

Figure D1. Measured (red) and simulated (blue) results of a hygrostatic controlled indoor climate (top: temperature, bottom: Relative humidity)

Context to the thesis

In this case, the same building is used as the one presented in Section 2.4, but the heating of the indoor air is now more advanced controlled. The validation results are quite satisfactory for this type of controller. Moreover, several control designs were simulated and evaluated

Acknowledgement

Edgar Neuhaus is acknowledged for his contribution

D2. Thermal comfort problems in a monumental office building in summer

From paper by H.L. Schellen , M. van Leth, M.A.P. van Aarle & A.W.M. van Schijndel
accepted for publication at 12 TH Building Physics Symposium Dresden 2007 March

One of the most important buildings in The Netherlands is the monumental building of the Senate. People working in the office rooms of this building have complaints on thermal comfort during summer time. A number of office rooms is overheated during warm summer days. Furthermore the rooms are ventilated in a natural way, i.e. by opening the windows. The installation of split air conditioning units or a HVAC system would have an unacceptable effect on the monumental interior and exterior. This paper will handle thermal comfort problems in a monumental office building in summer. One of the objectives of the work is to objectify the complaints. Furthermore it is the intention to have a better knowledge of the indoor climate of specific rooms in relation to the outdoor climate, their orientation and specific building physical properties and the use and related internal heat loads of the room. Moreover the aim of the work is to improve the summer indoor climate without affecting the monumental character of the rooms and the building itself. The method of approach is to objectify the complaints by measurements of the indoor climate in relation to the outdoor climate. Typical physical measurements considering thermal comfort were made. The results were compared with national guidelines regarding temperature exceeding limits, weighing hours and adaptive temperature limits. To improve the summer indoor conditions a simulation study on the indoor climate of a number of rooms was performed in HAMBASE (Heat, Air and Moisture, Buildings And Systems Engineering tool). The model was calibrated with the indoor climate results of the long term measurement sessions. A variant study on some improvement propositions was performed. Checking the indoor summer climate with the national guidelines indicated that about half of the measured rooms were too warm during warm summer days.

Exemplary results

Figure D2 shows the results of a model validation study and the use for design

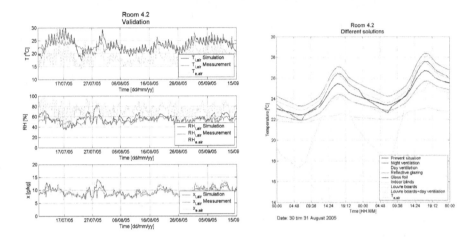

Figure D2 Left: Measured and simulated indoor climate of a warm office room Right: Simulated indoor air temperature for suggested (passive) measures

Context to the thesis

A whole building model is validated and used for the simulation and evaluation of several designs

Acknowledgement

Moniek van Leth is acknowledged for her contribution

D3. A hypocaust hot air floor heating system in the Netherlands

From paper by M.A.P. van Aarle, H.L. Schellen & A.W.M. van Schijndel accepted for publication at 12 TH Building Physics Symposium Dresden 2007 March

In 2002 a PhD study was finished on Heating Monumental Churches at the University of Technology in Eindhoven. Most of the used heating systems in the Netherlands were examined. However, at a number of places which were not accounted for in the previous mentioned study, unique heating systems are applied. The Stevens Church is heated by a hypocaust heating system: floor heating by hot air underneath the floor. Hot air is transported through a constructional duct system of bricks. A part of the hot air enters the church via a wall and floor air supplies, the rest is recirculated. The system is not very energy efficient: First, through the massive floor the heating system is very slow. It takes a very long time to heat up the church, almost 40 hours to heat up 7 °C. Second, air is not the most energetic medium for transport of heat. Third, a part of the capacity is used for heating up the crawl space and the ground. The purpose of this research is to design a more efficient heating system considering the preservation of monumental objects (such as church organs), the building itself and thermal comfort of the church attendance. The methodology was: (1) Measurement of the current indoor air temperatures, relative humidities, air inlet flows, air infiltration rate and external climate; (2) Simulation of the current indoor climate; (3) Validation by comparing measurements and simulations; (4) Simulation and evaluation of the design options given the specific criteria for indoor climate for the monumental objects and thermal comfort of the churchgoers. In the paper the results of previous mentioned methodology will be extensively discussed. It is concluded that a more optimal heating system for the Stevens Church would be a floor heating with warm water underneath the flags (stone) with additional hot air heating with floor air supplies.

Exemplary results

Figure D3 shows the results of a model validation study.

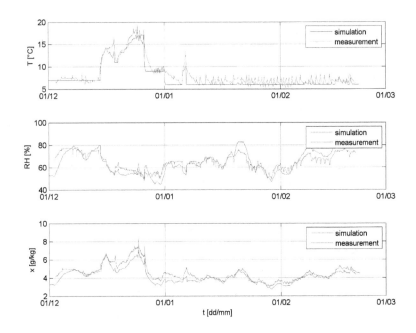

Figure D3 Measured and simulated of the current indoor climate of the church.

Context to the thesis

A whole building model is validated and used for the simulation and evaluation of several designs

Acknowledgement

Marcel van Aarle is acknowledged for his contribution

Appendix E

The thermal network of HAMBase revisited

In HAMBase, the view factors are approximated with the integrating sphere approach. The main advantages are twofold: First, there is no restriction on the geometrical form of the room and second, all walls are connected to a single temperature node (T_x) providing computational benefits. Section 2.2 provides a short summary of the model developed by de Wit [de Wit 1988;2006]. This appendix provides an additional example to better understand the thermal network presented in figure 2.2. The example comprehends a room with an air temperature Ta and two walls: a relative small wall with inner surface A_1 and temperature T_1 surrounded by the second wall (i.e. $A_2 \gg A_1$) with temperature T_2. Figure E1 shows the 'classical' thermal network including temperatures, heat flows and heat conductions.

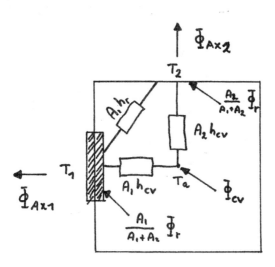

Figure E1. The thermal network of two wall nodes and one air node

In figure E1 both constructions are connected with the air node (T_a) and with each other by radiation. The net heat flows to the walls are respectively:

$$\Phi_{Ax1} = \frac{A_1}{A_1 + A_2}\Phi_r + A_1 h_{cv}(T_a - T_1) + A_1 h_r(T_2 - T_1) \tag{E1}$$

$$\Phi_{Ax2} = \frac{A_2}{A_1 + A_2}\Phi_r + A_2 h_{cv}(T_a - T_2) - A_1 h_r(T_2 - T_1) \tag{E2}$$

Figure E2 shows the HAMBase thermal model where the two walls are connected to a single node (T_x).

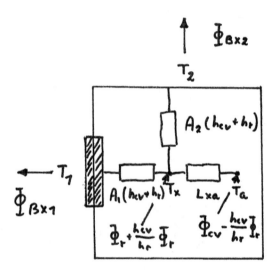

Figure E2. The HAMBase thermal network (see also figure 2.2)

In this case, the net heat flows to the walls are respectively:

$$\Phi_{Bx1} = A_1(h_{cv} + h_r)(T_x - T_1) \tag{E3}$$

$$\Phi_{Bx2} = A_2(h_{cv} + h_r)(T_x - T_2) \tag{E4}$$

The aim of this appendix is to verify that both thermal models of figure E1 and E2 are equivalent (assuming $A_1 \ll A_2$). This illustrates that the HAMBase thermal model in case of two walls is valid. A complete description is provided by [de Wit 2006].

We verify the equivalence of both models by showing that heat flows to the walls are approximate the same i.e. $\Phi_{Bx1} \approx \Phi_{Ax1}$ (a) and $\Phi_{Bx2} \approx \Phi_{Ax2}$ (b) and heat balances for the air nodes are the same in both cases (c).

The following equation is derived from the heat balance of node T_x of figure E2:

$$\Phi_r + \frac{h_{cv}}{h_r}\Phi_r = (T_x - T_1)A_1(h_{cv} + h_r) + (T_x - T_2)A_2(h_{cv} + h_r) + (T_x - T_a)L_{xa} \quad (E5)$$

Substituting Equation 2.1 of Chapter 2 i.e.:

$$L_{xa} = (A_1 + A_2)h_{cv}(\frac{h_r + h_{cv}}{h_r}) \quad (E6)$$

into E5 gives (after some basic algebra):

$$T_x = \frac{h_r A_1 T_1 + h_r A_2 T_2 + \Phi_r + h_{cv}(A_1 + A_2)T_a}{(A_1 + A_2)(h_r + h_{cv})} \quad (E7)$$

Now we can verify (a) by using equations E7, E3 and E1:

$$\Phi_{Bx1} = \frac{A_1}{A_1 + A_2}\Phi_r + A_1 h_{cv}(T_a - T_1) + A_1 h_r(\frac{A_1 T_1 + A_2 T_2}{A_1 + A_2} - T_1) =$$

$$\frac{A_1}{A_1 + A_2}\Phi_r + A_1 h_{cv}(T_a - T_1) + A_1 h_r(\frac{A_2 T_2}{A_1 + A_2} - (1 - \frac{A_1}{A_1 + A_2})T_1) \approx \quad (E8)$$

$$\frac{A_1}{A_1 + A_2}\Phi_r + A_1 h_{cv}(T_a - T_1) + A_1 h_r(T_2 - T_1) = \Phi_{Ax1}$$

195

We can verify (b) by using equations E7, E4 and E2:

$$\Phi_{Bx2} = \frac{A_2}{A_1 + A_2}\Phi_r + A_2 h_{cv}(T_a - T_2) + A_2 h_r(\frac{A_1 T_1 + A_2 T_2}{A_1 + A_2} - T_2) =$$

$$\frac{A_2}{A_1 + A_2}\Phi_r + A_2 h_{cv}(T_a - T_2) + \frac{A_2 h_r A_1 T_1}{A_1 + A_2} - \frac{A_2 h_r T_2 A_1}{A_1 + A_2}) \approx \qquad \text{(E9)}$$

$$\frac{A_2}{A_1 + A_2}\Phi_r + A_2 h_{cv}(T_a - T_2) - A_1 h_r(T_2 - T_1) = \Phi_{Ax2}$$

The last step is to show that the air node balances of both figures are the same. The heat balance air node T_a of figure E2 gives:

$$\Phi_{cv} - \frac{h_{cv}}{h_r}\Phi_r = L_{xa}(T_a - T_x) \qquad \text{(E10)}$$

using E6 and E7 gives (again after some basic algebra):

$$\Phi_{cv} - \frac{h_{cv}}{h_r}\Phi_r =$$

$$(A_1 + A_2)h_{cv}(\frac{h_r + h_{cv}}{h_r})T_a - \frac{h_{cv}}{h_r}(h_r A_1 T_1 + h_r A_2 T_2 + \Phi_r + h_{cv}(A_1 + A_2)T_a) \qquad \text{(E11)}$$

and so

$$\Phi_{cv} = (A_1 + A_2)h_{cv}T_a - h_{cv}A_1 T_1 - h_{cv}A_2 T_2 \qquad \text{(E12)}$$

The latter is the heat balance air node T_a of figure E1

Biography

Jos van Schijndel.

1987,   Ing. Physics Engineering at Fontys Polytechnical School, Eindhoven.

1991-1998,   Research Engineer at Technische Universiteit Eindhoven (TU/e).

1998,   MSc Physics at TU/e.

2007,   PhD. Building Physics at TU/e, PhD thesis entitled 'Integrated Heat Air and Moisture Modeling and Simulation'.

Since 1999   Assistant Professor for Building Physics and Systems Unit at TU/e.

Wissenschaftlicher Buchverlag bietet

kostenfreie

Publikation

von

wissenschaftlichen Arbeiten

Diplomarbeiten, Magisterarbeiten, Master und Bachelor Theses
sowie Dissertationen, Habilitationen und wissenschaftliche Monographien

Sie verfügen über eine wissenschaftliche Abschlußarbeit zu aktuellen oder zeitlosen
Fragestellungen, die hohen inhaltlichen und formalen Ansprüchen genügt,
und haben **Interesse an einer honorarvergüteten Publikation**?

Dann senden Sie bitte erste Informationen über Ihre Arbeit per Email
an info@vdm-verlag.de. Unser Außenlektorat meldet sich umgehend bei Ihnen.

VDM Verlag Dr. Müller Aktiengesellschaft & Co. KG
Dudweiler Landstraße 125a
D - 66123 Saarbrücken

www.vdm-verlag.de

www.ingramcontent.com/pod-product-compliance
Lightning Source LLC
LaVergne TN
LVHW022310060326
832902LV00020B/3380